Lecture Notes in Economics and Mathematical Systems   604

Donald Brown · Felix Kubler

Computational Aspects of General Equilibrium Theory

Refutable Theories of Value

Springer

Professor Donald Brown
Department of Economics
Yale University
27 Hillhouse Avenue
Room 15B
New Haven, CT 06520
USA
donald.brown@yale.edu

Professor Felix Kubler
Department of Economics
University of Pennsylvania
3718 Locust Walk
Philadelphia, PA 19104-6297
USA
fkubler@gmail.com

ISBN 978-3-540-76590-5   e-ISBN 978-3-540-76591-2

DOI 10.1007/978-3-540-76591-2

Lecture Notes in Economics and Mathematical Systems ISSN 0075-8442

Library of Congress Control Number: 2007939284

Production: LE-TEX Jelonek, Schmidt & Vöckler GbR, Leipzig
Cover design: WMX Design GmbH, Heidelberg

Printed on acid-free paper

9 8 7 6 5 4 3 2 1

springer.com

To Betty, Vanessa, Barbara and Elizabeth Rose
DJB

To Bi and He
FK

Preface

This manuscript was typeset in Latex by Mrs. Glena Ames. Mrs. Ames also drafted all the figures and edited the entire manuscript. Only academic custom prevents us from asking her to be a co-author. She has our heartfelt gratitude for her good humor and her dedication to excellence.

New Haven and Philadelphia,   *Donald J. Brown*
December 2007   *Felix Kubler*

Contents

List of Contributors

Donald J. Brown
Yale University
27 Hillhouse Avenue
New Haven, CT 06511
donald.brown@yale.edu

Caterina Calsamiglia
Universitat Autònoma de Barcelona
Edifici B
Bellaterra, Barcelona, Spain 08193
caterina.calsamiglia@uab.es

Ravi Kannan
Yale University
51 Prospect Street
New Haven, CT 06511
ravindran.kannan@yale.edu

Felix Kubler
University of Pennsylvania
3718 Locust Walk
Philadelphia, PA 19104-6297
kubler@sas.upenn.edu

Yoon-Ho Alex Lee
U.S. Securities & Exchange Commission
Washington, DC 20549
alex.lee@aya.yale.edu

Rosa L. Matzkin
Northwestern University
2001 Sheridan Road
Evanston, IL 60208
matzkin@northwestern.edu

Karl Schmedders
Northwestern University
2001 Sheridan Road
Evanston, IL 60208
k-schmedders@kellogg
.northwestern.edu

Chris Shannon
University of California at Berkeley
549 Evans Hall
Berkeley, CA 94720
cshannon@econ.berkeley.edu

Charles Steinhorn
Vassar College
124 Raymond Avenue
Poughkeepsie, NY 12604
steinhorn@vassar.edu

Refutable Theories of Value

Donald J. Brown[1] and Felix Kubler[2]

[1] Yale University, New Haven, CT 06511 `donald.brown@yale.edu`
[2] University of Pennsylvania, Philadelphia, PA 19104-6297 `kubler@sas.upenn.edu`

In the introduction to his classic *Foundations of Economic Analysis* [Sam47], Paul Samuelson defines meaningful theorems as "hypotheses about empirical data which could conceivably be refuted if only under ideal conditions." For three decades, the problems of existence, uniqueness and the stability of tâtonnement were at the core of the general equilibrium research program— see Blaug [Bla92], Ingaro and Israel [II90], and Weintraub [Wei85]. Are the theorems on existence, uniqueness and tâtonnement stability refutable propositions?

To this end, we define the Walrasian hypotheses about competitive markets:

H1. Market demand is the sum of demands of consumers derived from utility maximization subject to budget constraints at market prices.

H2. Market prices and consumer demands constitute a unique competitive equilibrium.

H3. Market prices are a locally stable equilibrium of the tâtonnement price adjustment mechanism.

The Walrasian model contains both theoretical constructs that cannot be observed such as utility and production functions and observable market data such as market prices, aggregate demand, expenditures of consumers or individual endowments. A meaningful theorem must have empirical implications in terms of observable market data.

In economic analysis there are two different methodologies for deriving refutable implications of theories. One method, used often in consumer theory and the theory of the firm, is marginal, comparative statics, and the other methodology is revealed preference theory. Both methods originated in Samuelson's *Foundations of Economic Analysis*.

We will follow the revealed preference approach. The proposition we shall need is Afriat's seminal theorem on the rationalization of individual consumer demand in competitive markets [Afr67]. Given a finite number of observations on market prices and individual consumer demands, his theorem states the equivalence between the following four conditions:

(a) The observations are consistent with maximization of a non-satiated utility function, subject to budget constraints at the market prices,
(b) There exists a finite set of utility levels and marginal utilities of income that, jointly with the market data, satisfy a set of inequalities called the Afriat inequalities,
(c) The observations satisfy a form of the strong axiom of revealed preference, involving only market data,
(d) The observations are consistent with maximization of a concave, monotone, continuous, non-satiated utility function, subject to budget constraints at the market prices.

The striking feature of Afriat's theorem is the equivalence of these four conditions. In particular, conditions (b) and (c). Moreover, condition (c) exhausts all refutable implications of a given data set unlike the necessary, but not sufficient, restrictions derivable from marginal, comparative statics.

The Afriat inequalities can be derived from the Kuhn–Tucker first-order conditions for maximizing a concave utility function subject to a budget constraint. These inequalities involve two types of variables: parameters and unknowns. Afriat assumes he can observe not only prices, but also individual demands. The other variables, utility levels, and marginal utilities of income are unknowns. But it follows from Afriat's theorem that the axiom in (c), a version of the strong axiom of revealed preference, containing only market data: prices and individual demands, is equivalent to the Afriat inequalities in (b) containing the unknowns: utility levels and marginal utilities of income. In going from (b) to (c), Afriat has managed to *eliminate all the unknowns*.

The Afriat inequalities are linear in the unknowns, if individual demands are observed. This is not the case when revealed preference theory is extended from individual demand to market demand, if individual demands are not observed. This nonlinearity in the Afriat inequalities is the major impediment in generalizing Afriat's and Samuelson's program on rationalizing individual demand in competitive markets to rationalizing market demand in competitive markets.

There are three general methods for deciding if a system of linear inequalities is solvable. The first method Fourier–Motzkin elimination, a generalization of the method of substitution taught in high school provides an exponential-time algorithm for solving a system of linear inequalities. In addition there are two types of polynomial-time algorithms for solving systems of linear inequalities: the ellipsoid method and the interior point method.

As an illustration of Fourier–Motzkin elimination, suppose that we have a finite set of linear inequalities in two real variables x and y. The solution set of this family of inequalities is a polyhedron in R^2. Applying the Fourier–Motzkin method to eliminate y amounts to projecting the points in the polyhedron onto the X-axis. Indeed, if x is in the projection, then we know there exists a y such that (x, y) is a point in the polyhedron defined by the set of linear inequalities. That is, (x, y) solves the system of linear inequalities.

If we carry out the Fourier–Motzkin elimination procedure, we can have three mutually exclusive outcomes. Either we discover that the inequalities are always satisfied: the projection is the whole X-axis, or there is no solution, i.e., the inequalities are inconsistent and the projection is empty. Or lastly, we are in the case we are most interested in, the case in which for some, but not all, values of x, the system has a solution: the projection is a nonempty proper subset of the X-axis. Fourier–Motzkin elimination is an instance of quantifier elimination. That is, we have eliminated the quantifier "there exists y."

In Brown and Matzkin [BM96], the logic of quantifier elimination is applied to analyze the refutability of H1. They derive a system of multivariate polynomial inequalities, where the unknowns are the utility levels, marginal utilities of income and individual demands in the Afriat inequalities and the individual demands in both the budget constraints and the aggregate demand conditions. The parameters are the market prices and the expenditures of consumers. This system of equilibrium inequalities is nonlinear in the unknowns; hence none of the methods cited above can be used to decide if the inequalities are solvable.

Brown and Matzkin show that H1 is refutable if and only if the family of equilibrium inequalities is refutable. That is, the system of inequalities is reduced to an equivalent system of multivariate polynomial inequalities in the parameters, where the system of multivariate polynomial inequalities is solvable iff the given parameter values satisfy the system of polynomial inequalities in the parameters. Moreover, the system of multivariate polynomials in the parameters exhausts all refutable implications of a given data set.

The Tarski–Seidenberg theorem [Tar51] on quantifier elimination provides an algorithm for deciding if a system of multivariate polynomial inequalities is refutable. This algorithm terminates in finite time in one of three mutually exclusive states: $(1 = 0)$, the given set of inequalities is never satisfied or $(1 = 1)$, the given set of inequalities in always satisfied or the system of inequalities is reduced to an equivalent system of multivariate polynomial inequalities in the parameters, where the system of multivariate polynomial inequalities is solvable if and only if the parameter values satisfy the system of polynomial inequalities in the parameters.

In our case, to argue that the algorithm cannot terminate with $1 = 0$, it is sufficient to invoke an existence theorem. But we actually do not need an existence theorem to conclude that the system of equilibrium inequalities is consistent, we only need an example where equilibrium exists. Similarly, to show that the algorithm cannot terminate with $1 = 1$, it suffices to construct an example, where in every equilibrium allocation some consumer's demands violate the revealed preference axiom in condition (c) of Afriat's theorem. Two such examples are given in Brown and Matzkin, proving that H1 is refutable.

Here is a simple example that illustrates quantifier elimination. Consider the quadratic equation $a(x^2) + bx + c = 0$. Here the unknown is x and the parameters are a, b and c. The equivalence between the existence of two real solutions to this equation and values of the parameters satisfying the inequality $(b^2) - 4ac > 0$ is an instance of quantifier elimination. Notice that the discriminant is only a function of the parameters, i.e., x, the unknown, has been eliminated.

A more interesting instance of quantifier elimination for economic analysis is the previously noted equivalence between conditions (b) and (c) in Afriat's theorem. To our knowledge, this is the first implicit application of quantifier elimination in economics.

Like Fourier–Motzkin elimination for linear inequalities, quantifier elimination is not a polynomial-time procedure. Fortunately, as mentioned above, to show refutability of the equilibrium inequalities we do not need to carry out quantifier elimination. It suffices to provide examples that rule out the $1 = 1$ and the $1 = 0$ states.

Of course, not all properties of an economic model need be refutable or meaningful in Samuelson's sense. This is evident from Afriat's theorem, where he shows that individual demand data is rationalizable, condition (a), if and only if it is rationalizable with a concave utility function, condition (d).

That is, the rationalization of individual demand data with concave utility functions cannot be refuted.

Brown and Shannon [BS00] show that H3 is not refutable. That is, in an exchange economy, if individual endowments are not observable and market equilibria are rationalizable as Walrasian equilibria then they can be rationalized as locally stable Walrasian equilibria under tâtonnement.

Finally, we consider H2. In an exchange economy, if individual endowments are not observable and market equilibria are rationablizable as Walrasian equilibria then it follows from the uniqueness of no-trade equilibria—see Balasko [Bal88]—that the market equilibria can be rationalized as unique Walrasian equilibria. That is, H2 is not refutable.

Many areas of applied economics such as finance, macroeconomics and industrial organization, use parametric equilibrium models to conduct counterfactual policy analysis. Often these models simply assume the existence of equilibrium. Moreover, it is difficult to determine the sensitivity of the policy analysis to the parametric specification.

This monograph presents a general equilibrium methodology for microeconomic policy analysis intended to serve as an alternative to the now classic, axiomatic general equilibrium theory as exposited in Debreu's *Theory of Value* [Deb59] or Arrow and Hahn's *General Competitive Analysis* [AH71].

The methodology proposed in this monograph does not presume the existence of market equilibrium, accepts the inherent indeterminacy of nonparametric general equilibrium models, offers effective algorithms for computing counterfactual equilibria in these models and extends Afriat's characterizations of individual supply and demand in competitive markets [Afr67, Afr72a] to aggregate supply and demand in competitive and non-competitive markets. The monograph consists of several essays that we have written over the past decade, some with colleagues or former graduate students, and an essay by Charles Steinhorn on the elements of O-minimal structures, the mathematical framework for our analysis.

The precursor to our research is Scarf's seminal *The Computation of Economic Equilibrium* [Sca73]. Scarf's algorithm uses a clever combinatorial argument to compute, in a finite number of iterations, an approximate fixed-point of any continuous map, f, of the simplex into itself. An approximate fixed-point is an x such that $f(x)$ differs in norm from zero by some given epsilon. In applied general equilibrium analysis, given parametric specifications of market supply and demand functions, Scarf's algorithm is used to compute market prices such that market demand and supply at these prices differ in norm by some given delta, constituting an approximate counterfactual equilibrium.

Recall that Arrow and Debreu's proof of existence of competitive equilibrium [AD54] and subsequent existence proofs, with the exception of Scarf's constructive argument, rely on Kakutani's fixed-point theorem [Kak41] or some other variant of Brouwer's fixed-point theorem [Bro10].

Surprisingly, Brouwer is also a founder of one of the schools of constructive analysis, Intuitionism. Brouwer at the end his career repudiated all mathematics that was non-constructive, i.e., proofs that invoke the law of the excluded middle, including his fixed-point theorem. In the school of constructive analysis created by Bishop—see his treatise, *Foundations of Constructive Analysis* [Bis67]—to prove the existence of a mathematical object requires a method or algorithm for constructing it.

The algorithmic or computational approach has displaced the axiomatic philosophy of Bourbaki, see Borel [Bor98], once dominant in contemporary mathematics and now common in economic theory after the publication of Debreu's *Theory of Value* [Deb59], Arrow's *Social Choice and Individual Values* [Arr51], and von Neumann and Morgenstern's *Theory of Games and Economic Behavior* [VM44].

The computational perspective has permeated allied fields of mathematics such as physics, chemistry and even biology. In this monograph we present nonparametric, computational theories of value that are, in principal, refutable by market data.

Policy analysis in applied general equilibrium theory requires a parametric specification of utility and production functions that are derived by calibration, where the specification is chosen to be consistent with one or more years of market data and estimated elasticities of market supply and demand—see Shoven and Whalley [SW92].

Here, our approach is nonparametric. Given a finite data set, we present an algorithm that constructs a semi-algebraic economy consistent with the data. That is, consumers' utility functions and firms' production functions are derived from market data using the Afriat inequalities—see Afriat [Afr67, Afr72a].

Semi-algebraic economies—economies where agent's characteristics such as utility functions and production functions are semi-algebraic, i.e., solutions of system multivariate polynomial inequalities—therefore arise naturally in refutable theories of value. Kubler and Schmedders [KS07] discuss the computation of Walrasian equilibria in semi-algebraic economies.

In general our models are indeterminate. This indeterminacy is minimized if data on individual consumption and production are available. The use of aggregate supply and demand data makes it difficult to numerically solve our models. Both Afriat [Afr67, Afr72a] and Varian [Var82, Var84] assumed observations on the consumption of households and the production of firms.

Under their assumptions our models reduce to a family of linear inequalities and a representative solution can be found using linear programming. We also compute representative solutions in the general case, but these algorithms are not polynomial-time algorithms as are interior-point linear programming algorithms.

An important aspect of the methodology presented here is a new class of existence theorems, where existence is conditional on the observed data. If the family of multivariate polynomial inequalities defining our model is solvable for the given data set, then there exist consumers and firms, i.e., utility functions and productions functions, such that the observed market prices are consistent with the behavioral assumptions and equilibrium notion of our model.

The classes of models where there are data sets for which the models are solvable and data sets where the models have no solution are called testable models. Brown and Matzkin [BM96] introduced the notion of testable model. In retrospect, a more descriptive term is refutable model.

It is important to note that Popper's notion of falsifiable scientific theories and our notion of refutable economic models are quite different concepts.

Refutability is a formal, deductive property of theoretical economic models just as identification is a formal, deductive property of econometric models. Identification is a necessary precondition for consistent estimation of an econometric model and refutability is a necessary precondition for falsification of a theoretical economic model i.e., subjecting the theory to critical empirical tests. Hence the joint hypotheses critique of falsification known as the Duhem–Quine thesis and other philosophical criticisms of the Popperian tradition in economics—see Part 2 in Backhouse [Bac94]—simply are not applicable to the notion of refutability.

The theory of revealed preference, due originally to Samuelson [Sam47] and culminating in the classic paper of Afriat [Afr67], may not be falsifiable in Popper's sense, but it is refutable in our sense, as is the Walrasian theory of general economic equilibrium—see the essay of Brown and Matzkin. That is, in both cases, these models when formulated in terms of the Afriat inequalities admit quantifier elimination of the unobserved variables. A result foreshadowed by a prescient letter from Poincaré to Walras in 1901, 50 years before Tarski's theorem on quantifier elimination.

Walras, in his attempts to persuade the French intellectual community to support his efforts to create a mathematical foundation for political economy, sent a copy of his magnum opus, *Elements of Pure Economics* [Wal54], to the most famous French mathematician of his day, Henri Poincaré. Walras had been severely criticized by French economists and mathematicians for claiming to derive observable, consumer demand functions from non-measurable, i.e., ordinal, utility functions.

In his reply to Walras' letter requesting an evaluation of his manuscript, Poincaré writes in part—"In your premises there are thus a certain number of arbitrary functions (ordinal utility functions): but as soon as you have laid down these premises, you have the right to draw consequences by means of calculus. If the arbitrary functions reappear in these consequences, the latter will not be false but devoid of all interest as subordinate to the arbitrary conventions set up at the beginning. You must therefore endeavor *to eliminate these arbitrary functions*, this is what you do."—see section 6.4 in Ingaro and Israel [II90].

Following Poincaré, this too is the methodological perspective of the essays in this monograph.

The essays in this monograph are on refutable economic models. In fact, we limit attention to semi-algebraic economic models. These are economic models defined by a finite family of multivariate polynomial inequalities and equations. The parameters in our models are derived from observable market data and the unknowns include unobservable theoretical constructs such as utility levels of consumers or marginal costs of firms, unobservable individual choices on the part of households and firms, and unobservable shocks to tastes or technology.

In the foundational monographs of Scarf, Debreu and Arrow and Hahn, cited above, there are three fundamental questions that must be answered by a theory of value:

1. Does the equilibrium exist?
2. Is the equilibrium Pareto optimal?
3. Can the equilibrium be effectively computed?

The essays by Brown and Matzkin [BM96], Brown and Shannon [BS00] and the first two essays by Kubler [Kub03] and [Kub04] are concerned with "existence." The essays by Brown and Calsamiglia [BC07] and Lee and Brown [LB07] are concerned with "optimality." The essay by Brown and Kannan [BK06] and the third essay by Kubler [Kub07] are concerned with "effective computation," as is the essay by Kubler and Schmedders [KS05].

Brown and Matzkin prove "existence" by applying Tarski's quantifier elimination algorithm [Tar51] to the finite family of polynomial inequalities and equations that define their model. This algorithm eliminates the unknowns from the model and terminates in a family of multivariate polynomial inequalities and equations over the parameters of the model.

If the parameter values given by the market data satisfy these inequalities and equations, then using Afriat's algorithm [Afr67, Afr72a] they construct utility functions and production functions such that the optimal consumption and production chosen by these agents, at the observed market prices, constitute a market equilibrium.

If the parameter values given by the market data do not satisfy these inequalities or equations then the model is refuted by the data. Consequently, refutable models can be used as specification tests for applied general equilibrium analysis.

Kubler's first two essays are concerned with refutability in dynamic stochastic models. A variation of Afriat's inequalities can be used to show that if all agents have time-separable expected utility and similar beliefs then competitive equilibrium imposes strong restrictions on the joint process of aggregate consumption and prices.

This suggests applications of our methodology to a wide variety of dynamic stochastic applied general equilibrium models. Production is not explicitly treated in Kubler's first essay, but its inclusion does not affect the results.

However, Kubler's second essay shows that the strong assumption of time-separable expected utility, commonly made in applied work, is crucial for the applicability of our methodology to dynamic stochastic models. If preferences are only assumed to be recursive and not necessarily representable as a time-separable expected utility, then the model cannot be refuted by market data even when individual choices are observed.

Quasilinear preferences are sufficient for partial equilibrium welfare analysis, where a consumer's welfare is measured by consumer surplus. Brown and Calsamiglia derive a revealed preference axiom for quasilinear preferences by eliminating the unknown utility levels from the Afriat inequalities under the assumption that the marginal utility of income is constant.

The Lee–Brown essay presents a general equilibrium welfare analysis of monopoly pricing, predatory pricing and mergers, despite the absence of a general equilibrium existence theorem. The counterfactual policy analysis presented in these essays is independent of the axiomatic, general equilibrium existence theorems that are a singular preoccupation of mathematical economics and general equilibrium theory—see Hildebrand's Introduction in Debreu [Deb86]. Consequently, refutable theories of value can address a broader spectrum of microeconomic policy issues than applied general equilibrium analysis, e.g., the Lee–Brown essay.

There are non-refutable—hence not falsifiable—properties of the semialgebraic, Walrasian theory of value. We have discussed at length the refutable implications of existence and now turn to the refutable implications of the Sonnenschein–Mantel–Debreu theorem [Deb74] on the stability of tâtonnement dynamics in a pure exchange economy.

Scarf [Sca60] was the first to construct an example of a pure exchange economy with a unique equilibrium that is globally unstable under tâtonnement dynamics. Subsequently, it was shown that any dynamic on the interior of the price simplex could be realized as tâtonnement dynamics for some pure exchange economy—the Sonnenschein–Mantel–Debreu theorem.

In the essay by Brown and Shannon, they show that both of these results are non-refutable or "not meaningful" theorems in Samuelson's sense. That is, if the Walrasian model can rationalize a market data set, then a Walrasian model can always rationalize it where the observed equilibria are locally stable under tâtonnement. Hence the class of nonparametric models considered in these essays enjoys properties not shared by parametric, applied general equilibrium models, where local stability of tâtonnement is problematic.

We now turn to computational issues. Tarski's algorithm or other quantifier elimination algorithms, such as Collin's cylindrical algebraic decomposition (CAD) algorithm [Col75] are not effective for our models. These algorithms may require an exponential number of iterations to eliminate all the unknowns. In this sense, solving our models is a "hard" problem. It is this algorithmic complexity that is addressed in the Brown–Kannan essay.

There is another literature on constructive or algorithmic analysis of the Walrasian theory of value, also inspired by Scarf's theorem on computing approximate fixed-points. The question motivating this literature is the effectiveness of algorithms for computing approximate fixed-points.

Richter and Wong [RW00] propose Turing machines as their definition of effective algorithms. Then they construct a self-map of the simplex, which has no (Turing) computable fixed points. It follows from Uzawa's [Uza62] theorem, on the equivalence of the Brouwer fixed point theorem and the existence of a competitive equilibrium in a pure exchange economy, that there is a pure exchange economy having no (Turing) computable equilibrium.

In the essay by Kubler and Schmedders [KS05], they examine how this can be reconciled with Scarf's notion of an approximate fixed point and his algorithm. They show that an approximate equilibrium can always be rationalized as an exact equilibrium of a 'close-by' economy and give various ways to formalize what 'close-by' means in dynamic stochastic economies.

Kubler and Schmedders [KS07] show that in a generic semi-algebraic economy, Walrasian equilibria are a subset of a finite set of solutions to a polynomial system of equations that can be derived from a finite set of polynomial inequalities and equations. There are several algorithms—see Sturmfels [Stu02]—to compute all solutions to polynomial systems.

Furthermore for such systems, Smale's alpha method [BCSS98] gives constructive, sufficient conditions for approximate zeros of the system of polynomial equations to be close to exact zeros. These approximate solutions to the system of multivariate polynomial inequalities and equations are close to the exact solutions of the system. Hence "almost is near" is a computationally effective notion in semi-algebraic economies and in Kubler and Schmedders [KS07].

Additional properties of semi-algebraic economies can be found in Blume and Zame [BZ93], where they extend Debreu's [Deb70] theorem on local uniqueness of equilibria in regular exchange economies to semi-algebraic exchange economies, and in Kubler and Schmedders [KS07].

Kubler's third essay extends the analysis of semi-algebraic economies, in the earlier essays, to O-minimal models. His essay uses Wilkie's theorem on O-minimal structures over Pfaffian functions — many economic models can be parameterized in terms of Pfaffian functions — to estimate the set of parameter values sufficient for computing counterfactuals in applied general equilibrium analysis.

These results on estimation of a set of parameter values provide an alternative to the parametric policy analysis described above and the fully nonparametric approach advocated in this monograph. Accepting the inherent indeterminacy of general equilibrium models, the parameter estimation results allow for approximate statements about parametric classes of O-minimal economies. An example in Kubler and Schmedders [KS07] is the estimation of the set of parameter values where an applied general equilibrium model has a unique equilibrium.

Charles Steinhorn is one of the original authors of the theory of O-minimal structures, a far reaching generalization of the theory of semi-algebraic sets, In the summer of 2003, at Don Brown's invitation, Charlie gave a series of lectures at the Cowles Foundation, intended for economists, on the elements of O-minimal structures. He has kindly consented to reprinting his lectures as an essay.

These lectures cover all the elements of O-minimal structures used in the body of the monograph. In particular, there is a proof of Tarski's theorem on quantifier elimination [Tar51] and a proof of Laskowski's theorem [Las92b] on the VC-property of a semi-algebraic family of sets, used in the Brown–Kannan algorithm for effectively computing counterfactual equilibria. Wilkie's theorem [Wil96] on Pfaffian functions is also discussed.

Charlie's lectures together with the mathematical and microeconomic prerequisites of standard graduate microeconomic texts, such as Varian [Var92] or Kreps [Kre90], suffice for reading these essays. Each essay is self-contained and they may be read in any order.

Acknowledgments

We would like to thank Rudy Bachmann for his careful reading and helpful critiques of previous versions of this essay.

Testable Restrictions on the Equilibrium Manifold

Donald J. Brown[1] and Rosa L. Matzkin[2]

[1] Yale University, New Haven, CT 06511 donald.brown@yale.edu
[2] Northwestern University, Evanston, IL 60208 matzkin@northwestern.edu

Summary. We present a finite system of polynomial inequalities in unobservable variables and market data that observations on market prices, individual incomes, and aggregate endowments must satisfy to be consistent with the equilibrium behavior of some pure trade economy. Quantifier elimination is used to derive testable restrictions on finite data sets for the pure trade model. A characterization of observations on aggregate endowments and market prices that are consistent with a Robinson Crusoe's economy is also provided.

Key words: General equilibrium, nonparametric restrictions, quantifier elimination, representative consumer

1 Introduction

The core of the general equilibrium research agenda has centered around questions on existence and uniqueness of competitive equilibria and stability of the price adjustment mechanism. Despite the resolution of these concerns, i.e., the existence theorem of Arrow and Debreu, Debreu's results on local uniqueness, Scarf's example of global instability of the tâtonnement price adjustment mechanism, and the Sonnenschein–Debreu–Mantel theorem, general equilibrium theory continues to suffer the criticism that it lacks falsifiable implications or in Samuelson' terms, "meaningful theorems."

Comparative statics is the primary source of testable restrictions in economic theory. This mode of analysis is most highly developed within the theory of the household and theory of the firm, e.g., Slutsky's equation, Shephard's lemma, etc. As is well known from the Sonnenschein–Debreu–Mantel theorem, the Slutsky restrictions on individual excess demand functions do not extend to market excess demand functions. In particular, utility maximization subject to a budget constraint imposes no testable restrictions on the set of equilibrium prices, as shown by Mas-Colell [Mas77]. The disappointing attempts of Walras, Hicks, and Samuelson to derive comparative statics for the general equilibrium model are chronicled in Inagro and Israel [II90]. Moreover, there has been no substantive progress in this field since Arrow and Hahn's discussion of monotone comparative statics for the Walrasian model [AH71].

If we denote the market excess demand function as $F_{\hat{w}}(p)$ where the profile of individual endowments $\hat{w}$ is fixed but market prices p may vary, then $F_{\hat{w}}(p)$ is the primary construct in the research on existence and uniqueness of competitive equilibria, the stability of the price adjustment mechanism, and comparative statics of the Walrasian model. A noteworthy exception is the monograph of Balasko [Bal88] who addressed these questions in terms of properties of the equilibrium manifold. To define the equilibrium manifold we denote the market excess demand function as $F(\hat{w}, p)$, where both $\hat{w}$ and p may vary. The equilibrium manifold is defined as the set $\{(\hat{w}, p) | F(\hat{w}, p) = 0\}$. Contrary to the result of Mas-Colell, cited above, we shall show that utility maximization subject to a budget constraint does impose testable restrictions on the equilibrium manifold.

To this end we consider an alternative source of testable restrictions within economic theory: the nonparametric analysis of revealed preference theory as developed by Samuelson, Houthakker, Afriat, Richter, Diewert, Varian, and others for the theory of the household and the theory of the firm. For us, the seminal proposition in this field is Afriat's theorem [Afr67], for data on prices and consumption bundles. Recall that Afriat, using the Theorem of the Alternative, proved the equivalence of a finite family of linear inequalities— now called the Afriat inequalities—that contain unobservable utility levels and marginal utilities of income with his axiom of revealed preference, "cyclical consistency"—finite families of linear inequalities that contain only observables (i.e., prices and consumption bundles), and with the existence of a concave, continuous monotonic utility function rationalizing the observed data. The equivalence of the Afriat inequalities and cyclical consistency is an instance of a deep theorem in model theory, the Tarski–Seidenberg theorem on quantifier elimination.

The Tarski–Seidenberg theorem—see van den Dries [Van88] for an extended discussion-proves that any finite system of polynomial inequalities can be reduced to an equivalent finite family of polynomial inequalities in the coefficients of the given system. They are equivalent in the sense that the original system of polynomial inequalities has a solution if and only if the parameter values of its coefficients satisfy the derived family of polynomial inequalities. In addition, the Tarski–Seidenberg theorem provides an algorithm which, in principle, can be used to carry out the elimination of the unobservable— the quantified—variables, in a finite number of steps. Each time a variable is eliminated, an equivalent system of polynomial inequalities is obtained, which contains all the variables except those that have been eliminated up to that point. The algorithm terminates in one of three mutually exclusive and exhaustive states: (i) $1 \equiv 0$, i.e., the original system of polynomial inequalities is never satisfied; (ii) $1 \equiv 1$, i.e., the original system is always satisfied; (iii) an equivalent finite family of polynomial inequalities in the coefficients of the original system which is satisfied only by some parameter values of the coefficients.

To apply the Tarski–Seidenberg theorem, we must first express the structural equilibrium conditions of the pure trade model as a finite family of polynomial inequalities. Moreover, to derive equivalent conditions on the data, the coefficients in this family of polynomial inequalities must be the market observables—in this case, individual endowments and market prices—and the unknowns must be the unobservables in the theory—in this case, individual utility levels, marginal utilities of income, and consumption bundles. A family of equilibrium conditions having these properties consists of the Afriat inequalities for each agent; the budget constraint of each agent; and the market clearing equations for each observation. Using the Tarski–Seidenberg procedure to eliminate the unknowns must therefore terminate in one of the following states: (i) $1 \equiv 0$—the given equilibrium conditions are inconsistent, (ii) $1 \equiv 1$—there is no finite data set that refutes the model, or (iii) the equilibrium conditions are testable.

Unlike Gaussian elimination-the analogous procedure for linear systems of equations-the running time of the Tarski–Seidenberg algorithm is in general not polynomial and in the worst case can be doubly exponential—see the volume edited by Arnon and Buchberger [AB88] for more discussion on the complexity of the Tarski–Seidenberg algorithm. Fortunately, it is often unnecessary to apply the Tarski–Seidenberg algorithm in determining if the given equilibrium theory has testable restrictions on finite data sets. It suffices to show that the algorithm cannot terminate with $1 \equiv 0$ or with $1 \equiv 1$. In fact, as we shall show, this is the case for the pure trade model.

It follows from the Arrow–Debreu existence theorem that the Tarski–Seidenberg algorithm applied to this system will not terminate with $1 \equiv 0$. In the next section, we construct an example of a pure trade model where no values of the unobservables are consistent with the values of the observables. Hence the algorithm will not terminate with $1 \equiv 1$. Therefore the Tarski–Seidenberg theorem implies for any finite family of profiles of individual endowments $\hat{w}$ and market prices p that these observations lie on the equilibrium manifold of a pure trade economy, for some family of concave, continuous, and monotonic utility functions, if and only if they satisfy the derived family of polynomial inequalities in $\hat{w}$ and p. This family of polynomial inequalities in the data constitute the testable restrictions of the Walrasian model of pure trade.

It may be difficult, using the Tarski–Seidenberg algorithm, to derive these testable restrictions on the equilibrium manifold in a computationally efficient manner for every finite data set, although we are able to derive restrictions for two observations. If there are more than two observations, our restrictions are necessary but not sufficient. That is, if our conditions hold for every pair of observations and there are at least three observations, then the data need not lie on any equilibrium manifold. Consequently, we call our conditions the weak axiom of revealed equilibrium or WARE. Of course, if our conditions are violated for any pair of observations, then the Walrasian model of pure trade is refuted.

An important distinction between our model and Afriat's model is we do not assume individual consumptions are observed as did Afriat. As a consequence the Afriat inequalities in our model are nonlinear in the unknowns.

This paper is organized as follows. Section 2 presents necessary and sufficient conditions for observations on market prices, individual incomes, and total endowments to lie on the equilibrium manifold of some pure trade economy. Section 3 specializes the results to equilibrium manifolds corresponding to economies whose consumers have homothetic utility functions. In the final section of the paper we discuss extensions and empirical applications of our methodology. In particular, we provide a characterization of the behavior of observations on aggregate endowments and market prices that is consistent with a Robinson Crusoe economy.

2 Restrictions in the Pure Trade Model

We consider an economy with K commodities and T traders, where the intended interpretation is the pure trade model. The commodity space is $\mathbb{R}^K$ and each agent has $\mathbb{R}^K_+$ as her consumption set. Each trader is characterized by an endowment vector $w_t \in \mathbb{R}^K_{++}$ and a utility function $V_t : \mathbb{R}^K_+ \to \mathbb{R}$. Utility functions are assumed to be continuous, monotone, and concave.

An *allocation* is a consumption vector x_t for each trader such that $x_t \in \mathbb{R}^K_+$ and $\sum_{t=1}^T x_t = \sum_{t=1}^T w_t$. The *price simplex* $\Delta = \{p \in \mathbb{R}^K_+ \,|\, \sum_{i=1}^K p_i = 1\}$. We shall restrict attention to strictly positive prices $S = \{p \in \Delta \,|\, p_i > 0 \text{ for all } i\}$. A *competitive equilibrium* consists of an allocation $\{x_t\}_{t-1}^T$ and prices p such that each x_t is utility maximizing for agent t subject to her budget constraint. The prices p are called *equilibrium prices*.

Suppose we observe a finite number N of profiles of individual endowment vectors $\{w_t^r\}_{t=1}^T$ and market prices p^r, where $r = 1, ..., N$, but we do not observe the utility functions or consumption vectors of individual agents. For each family of utility functions $\{V_t\}_{t=1}^T$ there is an equilibrium manifold, which is simply the graph of the Walras correspondence, i.e., the map from profiles of individual endowments to equilibrium prices.

We say that the pure trade model is *testable* if for every N there exists a finite family of polynomial inequalities in w_t^r and p^r for $t = 1, ..., T$ and $r = 1, ..., N$ such that observed pairs of profiles of individual endowments and market prices satisfy the given system of polynomial inequalities if and only if they lie on some equilibrium manifold.

To prove that the pure trade model is testable, we first recall Afriat's theorem [Afr67] (see also Varian [Var82]):

Theorem (Afriat's Theorem). *The following conditions are equivalent:*

(A.1) *There exists a nonsatiated utility function that "rationalizes" the data $(p^i, x^i)_{i=1,...,N}$; i.e., there exists a nonsatiated function $u(x)$ such that for all $i = 1, ..., N$, and all x such that $p^i \cdot x^i \geq p^i \cdot x$, $u(x^i) \geq u(x)$.*

(A.2) *The data satisfies "Cyclical Consistency (CC)" i.e. for all $\{r, s, t, ..., q\}$, $p^r \cdot x^r \geq p^r \cdot x^s$, $p^s \cdot x^s \geq p^s \cdot x^t$, ..., $p^q \cdot x^q \geq p^q \cdot x^r$ implies $p^r \cdot x^r = p^r \cdot x^s$, $p^s \cdot x^s = p^s \cdot x^t$, ..., $p^q \cdot x^q = p^q \cdot x^r$.*

(A.3) *There exist numbers U^i, $\lambda^i > 0$, $i = 1, ..., n$ such that $U^i \leq U^j + \lambda^j p^j \cdot (x^i - x^j)$ for $i, j = 1, ..., N$.*

(A.4) *There exists a nonsatiated, continuous, concave, monotonic utility function that rationalizes the data.*

Versions of Afriat's theorem for SARP (the Strong Axiom of Revealed Preference, due to Houthakker [Hou50]) and SSARP (the Strong SARP, due to Chiappori and Rochet [CR87]) can be found in Matzkin and Richter [MR91] and in Chiappori and Rochet [CR87], respectively.[3]

We consider the structural equilibrium conditions for N observations on pairs of profiles of individual endowment vectors $\{w_t^r\}_{t=1}^T$ and market prices p^r for $r = 1, ..., N$, which are:

$$\exists \{\bar{V}_t^r\}_{r=1,...,N;t=1,...,T}, \{\lambda_t^r\}_{r=1,...,N;t=1,...,T}, \{x_t^r\}_{r=1,...,N;t=1,...,T}$$

such that

$$\bar{V}_t^r - \bar{V}_t^s - \lambda_t^s p^s \cdot (x_t^s - x_t^s) \leq 0 \quad (r, s = 1, ..., N; \ t = 1, ..., T), \qquad (1)$$

$$\lambda_t^r > 0, \ x_t^r \geq 0 \quad (r = 1, ..., N; \ t = 1, ..., T), \qquad (2)$$

$$p^r \cdot x_t^r = p^r \cdot w_t^r \quad (r = 1, ..., N; t = 1, ..., T), \qquad (3)$$

$$\sum_{t=1}^T x_t^r = \sum_{t=1}^T w_t^r \quad (r = 1, ..., N). \qquad (4)$$

This family of conditions will be called the *equilibrium inequalities*. The observable variables in this system are the w_t^r and p^r, hence this is a nonlinear family of polynomial inequalities in unobservable utility levels, $\bar{V}_t^r$; marginal utilities of income, λ_t^r; and consumption vectors x_t^r. If we choose T concave, continuous and monotonic utility functions and N profiles of individual endowment vectors, then by the Arrow–Debreu existence theorem there exist equilibrium prices and competitive allocations such that the marginal utilities of income and utility levels of agents at the competitive allocations, together

---

[3] Chiappori and Rochet [CR87] show that SSARP characterizes demand data that can be rationalized by strictly monotone, strictly concave, C^∞ utility functions. Define the binary relationship R^0 by $x^t R^0 x$ if $p^t \cdot x^t \geq p^t \cdot x$. Let R be the transitive closure of R^0. Then, SARP is satisfied if and only if for all $t, s : [(x^t R x^s \ \& \ x^t \neq x^s) \Rightarrow (\text{not } x^s R x^t)]$; SSARP is SARP together with $[(p^s \neq \alpha p^r \text{ for all } \alpha = 0) \Rightarrow (x^s \neq x^r)]$.

with the competitive prices and allocations and profiles of endowment vectors, satisfy the equilibrium inequalities. Therefore, the Tarski–Seidenberg algorithm applied to the equilibrium inequalities will not terminate with $1 \equiv 0$.

The following example of a pure trade economy with two goods and two traders proves that the algorithm will not terminate with $1 \equiv 1$. In Figure 1, we superimpose two Edgeworth boxes, which are defined by the aggregate endowment vectors w^1 and w^2. The first box, (I), is $ABCD$ and the second box, (II), is $AEFG$. The first agent lives at the A vertex in both boxes and the second agent lives at vertex C in box (I) and at vertex F in box (II). The individual endowments $w^1_1, w^1_2; w^2_1, w^2_2$ and the two price vectors p^1 and p^2 define the budget sets of each consumer. The sections of the budget hyperplanes that intersect with each Edgeworth box are the set of potential equilibrium allocations. All pairs of allocations in box (I) and box (II) that lie on the given budget lines violate Cyclical Consistency for the first agent (the agent living at vertex A). By Afriat's theorem there is no solution to the equilibrium inequalities. This example is easily extended to pure trade models with any finite number of goods or traders.

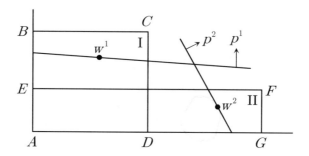

Fig. 1. Pure trade economy

Theorem 1. *The pure trade model is testable.*

Proof. The system of equilibrium inequalities is a finite family of polynomial inequalities; hence we can apply the Tarski–Seidenberg algorithm. We have shown above that the algorithm cannot terminate with $1 \equiv 0$ or with $1 \equiv 1$. □

It is often difficult to observe individual endowment vectors, so in the next theorem we restate the equilibrium inequalities where the observables are the market prices, incomes of consumers, and aggregate endowments. Let I^r_t denote the income of consumer t in observation r and w^r the aggregate endowment in observation r.

Theorem 2. *Let $\langle p^r, \{I_t^r\}_{t=1}^T, w^r\rangle$ for $r = 1, ..., N$ be given. Then there exists a set of continuous, concave, and monotone utility functions $\{V_t\}_{t=1}^T$ such that for each $r = 1, ..., N$: p^r is an equilibrium price vector for the exchange economy $\langle\{V_t\}_{t=1}^T, \{I_t^r\}_{t=1}^T, w^r\rangle$ if and only if there exists numbers $\{\bar{V}_t^r\}_{t=1,...,T; r=1,...,N}$ and $\{\lambda_t^r\}_{t=1,...,T; r=1,...,N}$ and vectors $\{x_t^r\}_{t=1,...,T; r=1,...,N}$ satisfying*

$$\bar{V}_t^r - \bar{V}_t^s - \lambda_t^s p^s \cdot (x_t^r - x_t^s) \quad (r, s = 1, ..., N; \ t = 1, ..., T), \tag{5}$$

$$\lambda_t^r > 0, \ x_t^r \geq 0 \quad (r = 1, ..., N; \ t = 1, ..., T), \tag{6}$$

$$p^r \cdot x_t^r = I_t^r \quad (r = 1, ..., N; \ t = 1, ..., T), \tag{7}$$

$$\sum_{t=1}^T x_t^r = w^r \quad (r = 1, ..., N). \tag{8}$$

Proof. Suppose that there exists $\{\bar{V}_t^r\}$, $\{\lambda_t^r\}$, and $\{x_t^r\}$ satisfying (5)–(8). Then, (5)–(7) imply, by Afriat's Theorem that for each t, there exists a continuous, concave, and monotone utility function $V_t : \mathbb{R}_+^K \to \mathbb{R}$ such that for each r, x_t^r is one of the maximizers of V_t subject to the budget constraint: $p^r y \leq I_t^r$. Hence, since $\{x_t^r\}_{t=1}^T$ define an allocation, i.e., satisfy (8), p^r is an equilibrium price vector for the exchange economy $\langle\{V_t\}_{t=1}^T, \{w_t^r\}_{t=1}^T\rangle$ for each $r = 1, ..., N$.

The converse is immediate, since given continuous, concave and monotone utility functions, V_t, the equilibrium price vectors p^r and allocations $\{x_t^r\}_{t=1}^T$ satisfy (7) and (8) by definition. The existence of $\{\lambda_t^r\}_{t=1}^T$ such that (5) and (6) hold follows from the Kuhn–Tucker Theorem, where $\bar{V}_t^r = V_t(x_t^r)$.   □

For two observations ($r = 1, 2$) and the Chiappori–Rochet version of Afriat's theorem we use, in the proof of Theorem 3 below, quantifier elimination to derive the testable restrictions for the pure trade model with two consumers ($t = a, b$) from the equilibrium inequalities. We call the family of polynomial inequalities obtained from this process the *Weak Axiom of Revealed Equilibrium* (WARE). To describe WARE, we let $\bar{z}_t^r$ ($r = 1, 2$; $t = a, b$) denote any vector such that $\bar{z}_t^r \in \arg\max_x\{p^s \cdot x | p^r \cdot x = I_t^r, \ 0 \leq x \leq w^r\}$ where $r \neq s$. Hence, among all the bundles that are feasible in observation r and are on the budget hyperplane of consumer t in observation r, $\bar{z}_t^r$ is any of the bundles that cost the most under prices p^s ($s \neq r$).

We will say that observations $\{p^r\}_{r=1,2}, \{I_t^r\}_{r=1,2;t=a,b}, \{w^r\}_{r=1,2}$ *satisfy* WARE if

(I) $\forall r = 1, 2, \ I_a^r + I_b^r = p^r \cdot w^r$,

(II) $\forall r, s = 1, 2 \ (r \neq s), \ \forall t = a, b, \ [(p^s \cdot \bar{z}_t^s \leq I_t^s) \Rightarrow (p^r \cdot \bar{z}_t^s > I_t^r)]$,

(III) $\forall r, s = 1, 2 \ (r \neq s), \ [(p^s \cdot \bar{z}_a^r \leq I_a^s) \& (p^s \cdot \bar{z}_b^r \leq I_b^s)] \Rightarrow (p^r \cdot w^s > p^r \cdot w^r)$.

In the next theorem we establish that WARE characterizes data that lie on some equilibrium manifold. Condition (I) says that the sum of the individuals' incomes equals the value of the aggregate endowment. Condition (II) applies when all the bundles in the budget hyperplane of consumer t in observation r that are feasible in observation r can be purchased with the income and prices faced by consumer t in observation s ($s \neq r$) (i.e., $p^s \cdot \bar{z}_t^r \leq I_t^s$). It says that it must then be the case that some of the bundles that are feasible in observation s and are in the budget hyperplane of consumer t in observation s cannot be purchased with the income and prices faced by consumer t in observation r (i.e., $p^r \cdot \bar{z}_t^s > I_t^r$). Clearly, unless this condition is satisfied, it will not be possible to find consumption bundles consistent with equilibrium and satisfying SSARP. Note that this condition is not satisfied by the observations in Figure 1. Condition (III) says that when for each of the agents it is the case that all the bundles that are feasible and affordable under observation r can be purchased with the agent's income and the price of observation s, then it must be that the aggregate endowment in observation s costs more than the aggregate endowment in observation r, with the prices of observation r. This guarantees that at least one of the pairs of consumption bundles in observation s that contain for each agent feasible and affordable bundles that could not be purchased with the income and price of observation r are such that they add up to the aggregate endowment.

Theorem 3. *Let $\{p^r\}_{r=1,2}$, $\{I_t^r\}_{r=1,2;t=a,b}$, $\{w^r\}_{r=1,2}$ be given such that p^1 is not a scalar multiple of p^2. Then the equilibrium inequalities for strictly monotone, strictly concave, C^∞ utility functions have a solution, i.e., the data lies on the equilibrium manifold of some economy whose consumers have strictly monotone, strictly concave, C^∞ utility functions, if and only if the data satisfy WARE.*

We provide in the Appendix a proof of Theorem 3 that uses the Tarski–Seidenberg theorem. A different type of proof is given in Brown and Matzkin [BM93].

3 Restrictions When Utility Functions Are Homothetic

In applied general equilibrium analysis—see Shoven and Whalley [SW92]—utility functions are often assumed to be homothetic. We next derive testable restrictions on the pure trade model under this assumption. These restrictions can be used as a specification test for computable general equilibrium models, say in international trade, where agents have homothetic utility functions.

Afriat [Afr77, Afr81] and Varian [Var83] developed the Homothetic Axiom of Revealed Preference (HARP), which is equivalent to the Afriat inequalities for homothetic utility functions. For two observations, $\{p^r, x^r\}_{r=1,2}$, HARP reduces to: $(p^r \cdot x^s)(p^s \cdot x^r) \geq (p^r \cdot x^r)(p^s \cdot x^s)$ for $r, s = 1, 2$ ($r \neq s$). If we substitute these for the Afriat inequalities in the equilibrium inequalities (1)–(4), we obtain a nonlinear system of polynomial inequalities where the unknowns (or unobservables) are the consumption vectors x_t^r for $r = 1, 2$ and $t = a, b$. Using quantifier elimination, we derive in the proof of Theorem 4 the testable restrictions of this model on the observable variables. We call these restrictions the *Homothetic-Weak Axiom of Revealed Preference (H-WARE)*.

Given observations $\{p^r\}_{r=1,2}$, $\{I_t^r\}_{r=1,2;t=a,b}$, $\{w^r\}_{r=1,2}$, we define the following terms:

$$\gamma_a = I_a^1 I_a^2, \quad \gamma_b = I_b^1 I_b^2, \quad \gamma_w = (p^1 \cdot w^2)(p^2 \cdot w^1),$$

$$\psi_1 = \gamma_b - \gamma_a - \gamma_w, \quad \psi_2 = (\gamma_b - \gamma_a - \gamma_w)^2 - 4\gamma_a \gamma_w,$$

$$r_1 = \frac{\gamma_a}{p^1 \bar{z}_a^2}, \quad r_2 = p^2 w^1 - \frac{\gamma_b}{p^1 \bar{z}_b^2},$$

$$t_1 = \frac{-\psi_1 - (\psi_2)^{1/2}}{2p^1 \cdot w^2}, \quad t_2 = \frac{-\psi_1 + (\psi_2)^{1/2}}{2p^1 \cdot w^2},$$

$$s_1 = \max\{r_1, t_1\}, \quad s_2 = \min\{r_2, t_2\}.$$

Let $\underline{z}_t^r$ ($r = 1, 2$; $t = a, b$) denote any vector such that $\underline{z}_t^r \in \arg\min_x \{p^s \cdot x \mid p^r \cdot x = I_t^r, \, 0 \le x \le w^r\}$ where $r \ne s$.

Our *Homothetic Weak Axiom of Revealed Equilibrium (H-WARE)* is

(H.I) $\Psi_2 \ge 0$,

(H.II) $s_1 \le s_2$,

(H.III) $s_1 \le p^2 \cdot \bar{z}_a^1$,

(H.IV) $I_a^1 + I_b^1 = p^1 \cdot w^1$ and $I_a^2 + I_b^2 = p^2 \cdot w^2$.

Condition (H.I) guarantees that t_1 and t_2 are real numbers. Conditions (H.II)–(H.IV) guarantee the existence of a vector x_a^1 whose cost under prices p^2 is between s_1 and s_2. The values of s_1 and s_2 guarantee that equilibrium allocations can be found. Condition (H.V) says that the sum of the individuals' incomes equals the value of the aggregate endowment.

Theorem 4. *Let $\{p^r\}_{r=1,2}$, $\{I_t^r\}_{r=1,2;t=a,b}$, $\{w^r\}_{r=1,2}$ be given. Then the equilibrium inequalities for homothetic utility functions have a solution, i.e., the data lie on the equilibrium manifold of some economy whose consumers have homothetic utility functions, if and only if the data satisfy H-WARE.*

In the Appendix, we provide a proof that uses the Tarski–Seidenberg theorem. See Brown and Matzkin [BM93] for a different proof.

4 Empirical Applications and Extensions

To empirically test the pure exchange model, one might use cross-sectional data to obtain the necessary variation in market prices and individual incomes. Assuming that sampled cities or states have the same distribution of tastes but different income distributions and consequently different market prices, the observations can serve as market data for our model. In the stylized economies in our examples one should think of each "trader" as an agent type, consisting of numerous small consumers, each having the same tastes and incomes.

There is a large variety of situations that fall into the structure of a general equilibrium exchange model and for which data are available. For example, our methods can be used in a multiperiod capital market model where agents have additively separable (time invariant) utility functions, to test whether spot prices are equilibrium prices, using only observations on the spot prices and the individual endowments in each period. They can be used to test the equilibrium hypothesis in an assets markets model where agents maximize indirect utility functions over feasible portfolios of assets, using observations on the outstanding shares of the assets, each trader's initial asset holdings, and the asset prices. Or, they can be used in a household labor supply model of the type considered in Chiappori [Chi88], to test whether the unobserved allocation of consumption within the household is determined by a competitive equilibrium, using data on the labor supply, wages, and the aggregate consumption of the household.

To apply the methodology to large data sets, it is necessary to devise a computationally efficient algorithm for solving large families of equilibrium inequalities. A promising approach is to restrict attention to special classes of utility functions. As an example, if traders are assumed to have quasilinear utility functions—all linear in the same commodity (say the kth)—then the equilibrium inequalities can be reduced to a family of linear inequalities by choosing the kth commodity as numeraire. We can now use the simplex algorithm or the interior point algorithm of Karmarkar—which runs in polynomial time—to test for or compute solutions of the equilibrium inequalities.

The more challenging problem in economic theory is to recast the equilibrium inequalities to allow random variation in tastes. Some recent progress has been made in this area by Brown and Matzkin [BM95]. They consider a random utility model, which gives rise to a stochastic family of Afriat inequalities, that can be identified and consistently estimated. If their approach can be extended to random exchange models then this is a significant step in empirically testing the Walrasian hypothesis.

The methodology can also be extended to find testable restrictions on the equilibrium manifold of economies with production technologies. Only observations on the market prices, individuals' endowments, and individuals' profit shares are necessary to test the equilibrium model in production economies. In particular, for a Robinson Crusoe economy, where the consumer has a nonsatiated utility function, we have derived the following restrictions on the observable variables, for any number of observations. A direct proof of the result is given in the Appendix.

Theorem 5. *The data $\langle p^r, w^r \rangle$ for $r = 1, ..., N$ lies in the equilibrium manifold of a Robinson Crusoe economy if and only if $\langle p^r, w^r \rangle$ for $r = 1, ..., N$ satisfy Cyclical Consistency (CC).*

Testable restrictions for other economic models can also be derived using the methodology that we have presented in this paper.

Acknowledgments

This is a revision of SITE Technical Report No. 85, "Walrasian Comparative Statistics," December 1993.

Support from NSF, Deutsche Fourschungsgemeinschaft, and Gottfried-Wilhelm-Leibnitz Forderpris is gratefully acknowledged. The first author wishes to thank the Miller Institute for its support. The second author wishes to thank the support of Yale University through a senior fellowship. This paper was written in part while the second author was visiting MIT, Princeton University, and the University of Chicago; their hospitality is gratefully acknowledged. We are indebted to Curtis Eaves, James Heckman, Daniel Mc-Fadden, Marcel Richter, Susan Snyder, Gautam Tripathi, and Hal Varian for helpful comments. We also thank participants in the various seminars and conferences at which previous versions of this paper were presented for their remarks. Comments of the editor and the referees have greatly improved the exposition in this paper. The typing assistance of Debbie Johnston is much appreciated.

Brown, D.J., Matzkin, R.L.: Testable restrictions on the equilibrium manifold. Econometrica **64**, 1249–1262(1996). Reprinted by permission of the Econometric Society.

Appendix

Proof of Theorem 3 Using the Tarski–Seidenberg theorem, we need to show that WARE can be derived by quantifier elimination from the equilibrium inequalities for strictly monotone, strictly concave, C^∞ utility functions. Making use of Chiappori and Rochet [CR87], these inequalities are: $\exists \{\bar{V}_t^r\}_{r=1,2;t=a,b}$, $\{\lambda_t^r\}_{r=1,2;t=a,b}$, $\{x_t^r\}_{r=1,2;t=a,b}$ such that

(C.1) $\bar{V}_t^2 - \bar{V}_t^1 - \lambda_t^1 p^1 \cdot (x_t^2 - x_t^1) < 0, \quad t = a, b;$

(C.2) $\bar{V}_t^1 - \bar{V}_t^2 - \lambda_2^t p^2 \cdot (x_t^1 - x_t^2) < 0, \quad t = a, b;$

(C.3) $\lambda_t^r > 0, \quad r = 1, 2; \ t = a, b;$

(C.4) $p^r \cdot x_t^r = I_t^r, \quad r = 1, 2; \ t = a, b;$

(C.5) $p^1 \neq p^2 \Rightarrow x_t^1 \neq x_t^2, \quad t = a, b;$

(C.6) $x_t^r \geq 0, \quad r = 1, 2; \quad t = a, b;$

(C.7) $a^r + x_b^t = w^r, \quad r = 1, 2.$

The equivalent expression, after eliminating $\{\lambda_t^r\}_{r=1,2;t=a,b}$, is $\exists \{\bar{V}_t^r\}_{r=1,2;t=a,b}$, $\{x_t^r\}_{r=1,2;t=a,b}$ such that

(C.1′) $p^1 \cdot (x_t^2 - x_t^1) \leq 0 \Rightarrow \bar{V}_t^2 < \bar{V}_t^1, \quad t = a, b;$

(C.2′) $p^2 \cdot (x_t^1 - x_t^2) \leq 0 \Rightarrow \bar{V}_t^1 < \bar{V}_t^2, \quad t = a, b;$

(C.4) $p^r \cdot x_t^r = I_t^r, \quad r = 1, 2; \quad t = a, b;$

(C.5) $p^1 \neq p^2 \Rightarrow x_t^1 \neq x_t^2, \quad t = a, b;$

(C.6) $x_t^r \geq 0, \quad r = 1, 2; \quad t = a, b;$

(C.7) $x_a^r + x_b^t = w^r, \quad r = 1, 2.$

Necessity is clear. Sufficiency follows by noticing that (C.1′) and (C.2′) imply, respectively, that $\exists \{\lambda_t^1\}_{t=a,b}$ satisfying (C.1) and (C.3) and $\exists \{\lambda_t^2\}_{t=a,b}$ satisfying (C.2) and (C.3). Elimination of $\{\bar{V}_t^r\}_{t=1,2;t=a,b}$ yields the equivalent expression: $\exists \{x_t^r\}_{r=1,2;t=a,b}$ such that

(C.1″) $p^1 \cdot (x_t^2 - x_t^1) \leq 0 \Rightarrow p^2 \cdot (x_t^1 - x_t^2) > 0, \quad t = a, b;$

(C.4) $p^r \cdot x_t^r = I_t^r, \quad r = 1, 2; \quad t = a, b;$

(C.5) $p^1 \neq p^2 \Rightarrow x_t^1 \neq x_t^2, \quad t = a, b;$

(C.6) $x_t^r \geq 0, \quad r = 1, 2; \quad t = a, b;$

(C.7) $x_a^r + x_b^t = w^r, \quad r = 1, 2.$

This follows because (C.1″) is necessary and sufficient for the existence of $\{\bar{V}_t^r\}_{r=1,2;t=a,b}$ satisfying (C.1′)–(C.2′). Note that we have just shown how, for two observations, SSARP can be derived by quantifier elimination. Next, elimination of $\{x_b^r\}_{t=1,2}$, using (C.7), yields the equivalent expression: $\exists x_a^1, x_a^b$ such that

(C.1‴.1) $p^2 \cdot x_a^1 \leq I_a^2 \Rightarrow p^1 \cdot x_a^2 > I_a^1;$

(C.2‴.2) $p^2 \cdot (w^1 - x_a^1) \leq I_b^2 \Rightarrow p^1 \cdot (w^2 - x_a^2) > I_b^1;$

(C.4′) $\quad p^r \cdot x_a^r - I_a^r, \quad r = 1, 2;$

(C.5′) $\quad p^1 \neq p^2 \Rightarrow [(x_a^1 \neq x_a^2) \ \& \ (w^1 - x_a^1 \neq w^2 - x_a^2)];$

(C.6′) $\quad 0 \leq x_a^r \leq w^r, \quad r = 1, 2;$

(C.7′) $\quad I_a^r + I_b^r = p^r \cdot w^r, \quad r = 1, 2.$

Let $\underline{z}_t^r$ denote any vector such that $\underline{z}_t^r \in \arg\min_x \{p^s \cdot x \mid p^r \cdot x = I_t^r, 0 \leq x \leq w^r\}$, where $r \neq s$. Then, after elimination of x_a^2 we get: $\exists x_a^1$ such that

(C.1⁗.1) $p^2 \cdot x_a^1 \leq I_a^2 \Rightarrow p^1 \cdot x_a^2 > I_a^1;$

(C.2⁗.2) $p^2 \cdot (w^1 - x_a^1) \leq I_b^2 \Rightarrow p^1 \cdot (w^2 - \underline{z}_a^2) > I_b^1;$

(C.3''''.3) $[(p^2 x_a^1 \le I_a^2) \ \& \ (p^2 \cdot (w^1 - x_a^1) \le I_b^2)] \Rightarrow p^1 \cdot w^2 > p^2 \cdot w^1;$

(C.4') $p^1 \cdot x_a^1 = I_a^1;$

(C.6') $0 \le x_a^1 \le w^1, \quad r = 1, 2;$

(C.7') $I_a^r + I_b^r = p^r \cdot w^r, \quad r = 1, 2.$

Necessity of (C.1''''.1) and (C.1''''.2) follows by the definitions of $\underline{z}_a^2 2$; and $\underline{z}_b^2$. Necessity of (C.1''''.3) follows by using (C.1'''.1), (C.1'''.2), and (C.7'). The existence of x_a^1 satisfying (C.1'''.1), (C.1'''.2). (C.4')–(C.7') follows immediately if $(p^2 \cdot x_a^1 > I_a^2) \ \& \ (p^2 \cdot (w^1 - x_a^1) > I_b^2)$; it follows using (C.1''''.1) if $(p^2 \cdot x_a^1 \le I_a^2) \ \& \ (p^2 \cdot (w^1 - x_a^1) > I_b^2)$; it follows using (C.1''''.2) if $(p^2 \cdot x_a^1 > I_a^2) \ \& \ (p^2 \cdot (w^1 - x_a^1) \le I_b^2)$; and it follows using (C.1''''.1)–(C.1''''.3) if $(p^2 \cdot x_a^1 \le I_a^2) \ \& \ (p^2 \cdot (w^1 - x_a^1) \le I_b^2)$. (C.5') can always be satisfied. Finally, elimination of x_a^1 yields, by similar arguments, the equivalent expression:

(C.1*.1) $p^1 \times \bar{z}_a^2 \le I_a^1 \Rightarrow p^2 \cdot \bar{z}_a^1 > I_a^2;$

(C.1*.2) $p^1 \cdot (w^2 - \underline{z}_a^2) \le I_a^1 \Rightarrow p^2 \cdot (w^1 - \underline{z}_a^2) > I_a^2;$

(C.1*.3) $[(p^1 \cdot \bar{z}_a^2 \le I_a^1) \ \& \ (p^1 \cdot (w^2 - \underline{z}_a^2) \le I_b^1)] \Rightarrow p^2 \cdot w^1 > p^2 \cdot w^2;$

(C.1*.4) $[(p2 \cdot \bar{z}_a^1 \le I_a^2) \ \& \ (p^2 \cdot (w^1 - \underline{z}_a^1 1) \le I_b^2)] \Rightarrow p^1 \cdot w^2 > p^1 \cdot w^1;$

(C.7') $I_a^r + I_b^r = p^r \cdot w^r, \quad r = 1, 2.$

Note that $I_a^r + I_b^r = p^r \cdot w^r$ implies that $p^s \cdot \bar{z}_a^r + p^s \cdot \underline{z}_b^r = p^s \cdot w^r \ (s \ne r)$. Hence, the above family of polynomial inequalities can be written as:

(I) $\forall r = 1, 2, \ I_a^r + I_b^r = p^r \cdot w^r;$

(II) $\forall r, s = 1, 2 (r \ne s), \forall t = a, b, [(p^s \cdot \bar{z}_t^r \le I_t^s) \Rightarrow (p^r \cdot \bar{z}_t^s > I_t^r)];$

(III) $\forall r, s = 1, 2 (r \ne s), [(p^s \cdot \bar{z}_a^r \le I_a^s) \& (p^s \cdot \bar{z}_b^r \le I_b^s)] \Rightarrow (p^r \cdot w^r > p^r \cdot w^r)$

which is our *Weak Axiom of Revealed Equilibrium* (WARE).

Proof of Theorem 4. Using the Tarski–Seidenberg theorem, we show that H-WARE can be derived by quantifier elimination from the equilibrium inequalities for homothetic, concave, and monotone utility functions, Hence, we have to eliminate the quantifiers in the following expression: $\exists x_a^1, x_a^2, x_b^1, x_b^2$ such that

(H.1) $(p^1 \cdot x_a^2)(p^2 \cdot x_a^1) \ge \gamma_a;$

(H.2) $(p^1 \cdot x_b^2)(p^2 \cdot x_b^1) \ge \gamma_b;$

(H.3) $p^r \cdot x_t^r = I_t^r, \ r = 1, 2; \ t = a, b;$

(H.4) $x_t^r \ge 0, \ r = 1, 2; \ t = a, b;$

(H.5) $x_a^r + x_b^r = w^r, \ r = 1, 2.$

This is equivalent to: $\exists x_a^1, x_a^2$ such that

(H.1) $(p^1 \cdot x_a^2)(p^2 \cdot x_a^1) \ge \gamma_a;$

(H.2′) $(p^1 \cdot (w^2 - x_a^2))(p^2 \cdot (w^1 - x_a^1)) \geq \gamma_b'$;

(H.3′) $p^r \cdot x_a^r = I_a^r$, $r = 1, 2$;

(H.4′) $w^r \geq x_a^r \geq 0$, $r = 1, 2$;

(H.5′) $I_a^r + I_b^r = p^r \cdot w^r$, $r = 1, 2$.

(H.1) and (H.2′) can be expressed as:

(H.1′) $p^1 \cdot w^2 - \dfrac{\gamma_b}{p^2 \cdot (w^1 - x_a^1)} \geq p^1 \cdot x_a^2 \geq \dfrac{\gamma_a}{p^2 \cdot x_a^1}$.

So, the expression: "$\exists x_a^1, x_a^2$ satisfying (H.1′)–(H.5′)" is equivalent to: $\exists x_a^1$ such that

(H.1.1) $p^1 \cdot \bar{z}_a^2 \geq \gamma_a p^2 \cdot x_a^1$;

(H.1.2) $p^1 \cdot w^2 - \dfrac{\gamma_b}{p^2 \cdot (w^1 - x_a^1)} \geq p^1 \cdot \bar{z}_a^2$;

(H.1.3) $p^1 \cdot w^2 - \dfrac{\gamma_b}{p^2 \cdot (w^1 - x_a^1)} \geq \dfrac{\gamma_a}{p^2 \cdot x_a^1}$;

(H.3″) $p^1 \cdot x_a^1 = I_a^1$, $r = 1, 2$;

(H.4″) $w^1 \geq x_a^1 \geq 0$, $r = 1, 2$;

(H.5′) $I_a^r + I_b^r = p^r \cdot w^r$, $r = 1, 2$;

or, equivalently, to: $\exists x_a^1$ such that

(H.1.1′) $\dfrac{\gamma_w - (p^1 \cdot \bar{z}_a^2)(p^2 \cdot w^1) - \gamma_b}{(p^1 \cdot w^2 - p^1 \cdot \bar{z}_a^2)} \geq p^2 \cdot x_a^1 \geq \dfrac{\gamma_a}{p^1 \cdot \bar{z}_a^2}$;

(H.1.2′) $(p^1 \cdot w^2)(p^2 \cdot x_a^1)^2 + (\gamma_b - \gamma_w - \gamma_a)(p^2 \cdot x_a^1) + \gamma_a p^2 \cdot w^1 \leq 0$;

(H.3″) $p^1 \cdot x_a^1 = I_a^1$, $r = 1, 2$;

(H.4″) $w^1 \geq x_a^1 \geq 0$, $r = 1, 2$;

(H.5′) $I_a^r + I_b^r = p^r \cdot w^r$, $r = 1, 2$.

Using the fact that $p^1 \cdot (w^2 - \bar{z}_a^2) = p^1 \cdot \bar{z}_b^2$, (H.1.1′) can be written as

$$p^2 \cdot w^1 - \dfrac{\gamma_b}{p^1 \cdot \bar{z}_b^2} \geq p^2 \cdot x_a^1 \geq \dfrac{\gamma_a}{p^1 \cdot \bar{z}_a^2},$$

or, equivalently, as

(H.1.1″) $r_2 \geq p^2 \cdot x_a^1 \geq r_1$.

The necessary and sufficient conditions for the existence of x_a^1 satisfying (H.1.1″), (H.1.2′), (H.3″), (H.4″), (H.5′) are

(H.1*) $r_1 \leq p^2 \cdot \bar{z}_a^1$, $p^2 \cdot \underline{z}_a^1 \leq r_2$, $r_1 \leq r_2$;

(H.2*) $\Psi_2 = (\Psi_1)^2 - 4\gamma_a \gamma_w \geq 0$;

(H.3*) $t_1 \leq p^2 \cdot \bar{z}_a^1$, $p^2 \cdot \underline{z}_a^1 \leq t_2$;

(H.4*) $I_a^r + I_b^r = p^2 \cdot w^r$, $r = 1, 2$;

or, equivalently, the conditions are

(H.I) $\Psi_2 \geq 0$;

(H.II) $s_1 \leq s_2$;

(H.III) $s_1 \leq p^2 \cdot \bar{z}_a^1$;

(H.IV) $p^2 \cdot \underline{z}_a^1 \leq s_2$;

(H.V) $I_a^1 + I_b^1 = p^1 \cdot w^1$ and $I_a^2 + I_b^2 = p^2 \cdot w^2$;

which is our Homothetic Axiom of Revealed Preference. Necessity is clear. To show sufficiency, note that (H.1)–(H.IV) imply that $\exists x_a^1$ satisfying (H.3″)–(H.4″) and $\max\{r_1, t_1\} \leq p^2 \cdot x_a^1 \leq \min\{r_1, t_2\}$. That such x_a^1 satisfies (H.1.1″) is obvious. That it satisfies (H.1.2′) follows because the function $f(t) = (t - t_1)(t - t_2)$ is such that $f(t) \leq 0$ for all $t \in [t_1, t_2]$ and (H.1.2′) can be written as $(p^2 \cdot x_a^1 - t_1)(p^2 \cdot x_a^1 - t_2) \leq 0$. $\qquad\square$

Proof of Theorem 5. Let x^r and y^r denote, respectively, a consumption and production plan in observation r. If $\langle p^r, w^r \rangle_{r=1}^N$ satisfy CC, then $\langle p^r, x^r = w^r, y^r = 0 \rangle_{r=1,...,N}$ satisfy the Afriat inequalities for utility maximization and profit maximization (see Varian [Var84]), and markets clear. Suppose that $\langle p^r, w^r \rangle_{r=1}^N$ does not satisfy CC but lies in the equilibrium manifold. Let x^r and y^r denote, respectively, any equilibrium consumption and equilibrium production plan in observation r. Since CC is violated, there exists $\{s, v, f, ..., e\}$ such that

$$p^s \cdot w^v \leq p^s \cdot w^s, p^v \cdot w^f \leq p^v \cdot w^v, ..., p^e \cdot w^s \leq p^e \cdot w^e \qquad (9)$$

where at least one of the inequalities is strict. Profit maximization ($p^s \cdot y^v \leq p^s \cdot y^s, p^v \cdot y^f \leq p^v \cdot y^v, ..., p^e \cdot y^s \leq p^e \cdot y^e$) and markets clearing ($x^v = w^v + y^v, x^s = w^s + y^s, x^f = w^f + y^f, ..., x^e = w^e + y^e$) imply with (9) that

$$p^s \cdot x^v \leq p^s \cdot x^s, p^v \cdot x^f \leq p^v \cdot x^v, ..., p^e \cdot x^s \leq p^e \cdot x^e \qquad (10)$$

where at least one of the inequalities is strict. Since (10) is inconsistent with utility maximization, a contradiction has been found.

Uniqueness, Stability, and Comparative Statics in Rationalizable Walrasian Markets

Donald J. Brown[1] and Chris Shannon[2]

[1] Yale University, New Haven, CT 06520-8269 donald.brown@yale.edu
[2] University of California at Berkeley, Berkeley, CA 94720
   cshannon@econ.berkeley.edu

Summary. This paper studies the extent to which qualitative features of Walrasian equilibria are refutable given a finite data set. In particular, we consider the hypothesis that the observed data are Walrasian equilibria in which each price vector is locally stable under tâtonnement. Our main result shows that a finite set of observations of prices, individual incomes and aggregate consumption vectors is rationalizable in an economy with smooth characteristics if and only if it is rationalizable in an economy in which each observed price vector is locally unique and stable under ttonnement. Moreover, the equilibrium correspondence is locally monotone in a neighborhood of each observed equilibrium in these economies. Thus the hypotheses that equilibria are locally stable under tâtonnement, equilibrium prices are locally unique and equilibrium comparative statics are locally monotone are not refutable with a finite data set.

Key words: Local stability, Monotone demand, Refutability, Equilibrium Manifold

1 Introduction

The major theoretical questions concerning competitive equilibria in the classical Arrow–Debreu model—existence, uniqueness, comparative statics, and stability of price adjustment processes—have been largely resolved over the last forty years. With the exception of existence, however, this resolution has been fundamentally negative. The conditions under which equilibria can be shown to be unique, comparative statics globally determinate or tâtonnement price adjustment globally stable are quite restrictive. Moreover, the Sonnenschein–Debreu–Mantel theorem shows in striking fashion that no behavior implied by individual utility maximization beyond homogeneity and Walras' Law is necessarily preserved by aggregation in market excess demand. This arbitrariness of excess demand implies that monotone equilibrium comparative statics and global stability of equilibria under tâtonnement will only result from the imposition of a limited set of conditions on preferences and endowments. Based on these results, many economists conclude that the general equilibrium model has no refutable implications or empirical content.

Of course no statement concerning refutable implications is meaningful without first specifying what information is observable and what is unobservable. If only market prices are observable, and all other information about the economy such as individual incomes, individual demands, individual endowments, individual preferences, and aggregate endowment or aggregate consumption is unobservable, then indeed the general equilibrium model has no testable restrictions. This is essentially the content of Mas-Colell's version of the Sonnenschein–Debreu–Mantel theorem. Mas-Colell [Mas77] shows that given an arbitrary nonempty compact subset C of strictly positive prices in the simplex, there exists an economy $\mathcal{E}$ composed of consumers with continuous, monotone, strictly convex preferences such that the equilibrium price vectors for the economy $\mathcal{E}$ are given exactly by the set C. In many instances, however, it is unreasonable to think that only market prices are observable; other information such as individual incomes and aggregate consumption may be observable in addition to market prices. Brown and Matzkin [BM96] show that if such additional information is available, then the Walrasian model does have refutable implications. They demonstrate by example that with a finite number of observations—in fact two—on market prices, individual incomes and aggregate consumptions, the hypothesis that these data correspond to competitive equilibrium observations can be rejected. They also give conditions under which this hypothesis is accepted and there exists an economy rationalizing the observed data.[3]

This paper considers the extent to which qualitative features of Walrasian equilibria are refutable given a finite data set. In particular, we consider the hypothesis that the observed data are Walrasian equilibria in which each price vector is locally stable under tâtonnement. Based on the Sonnenschein–Debreu–Mantel results and the well-known examples of global instability of tâtonnement such as Scarf's [Sca60], it may seem at first glance that this hypothesis will be easily refuted in a Walrasian setting. Surprisingly, however, we show that it is not. Our main result shows that a finite set of observations of prices, individual incomes and aggregate consumption vectors is rationalizable in an economy with smooth characteristics if and only if it is rationalizable in a distribution economy in which each observed price is locally stable under tâtonnement. Moreover, the equilibrium correspondence is locally monotone in a neighborhood of each observed equilibrium in these economies, and the equilibrium price vector is locally unique.

The conclusion that if the data is rationalizable then it is rationalizable in a distribution economy, i.e., an economy in which individual endowments are collinear, is not subtle. If we do not observe the individual endowments

---

[3] Recent work by Chiappori and Ekeland [CE98] considers a related question. They show that observations of aggregate endowments and prices place no restrictions on the local structure of the equilibrium manifold.

and only observe prices and income levels, then one set of individual endowments consistent with this data is collinear, with shares given by the observed income distribution. Since distribution economies by definition have a price-independent income distribution, this observation may suggest that our results about stability and comparative statics derive simply from this fact. Kirman and Koch [KK86] show that this intuition is false, however. They show that the additional assumption of a fixed income distribution places no restrictions on excess demand: given any compact set $K \subset \mathbb{R}^{\ell}_{++}$ and any smooth function $g : \mathbb{R}^{\ell}_{++} \to \mathbb{R}$ which is homogeneous of degree 0 and satisfies Walras' Law, and given any fixed income distribution $\alpha \in \mathbb{R}^n_{++}$, $\sum_{i=1}^{n} \alpha_i = 1$, there exists an economy $\mathcal{E}$ with smooth, monotone, strictly convex preferences and initial endowments $\omega_t = \alpha_t \omega$ such that excess demand for $\mathcal{E}$ coincides with g on K. Hence any dynamic on the price simplex can be replicated by some distribution economy.

This paper shows that rationalizable data is always rationalizable in an economy in which market excess demand has a very particular structure. Using recent results of Quah [Qua98], we show that if the data is rationalizable then it is rationalizable in an economy in which each individual demand function is locally monotone at each observation. The strong properties of local monotonicity, in particular the fact that local monotonicity of individual demand is preserved by aggregation in market excess demand and the fact that local monotonicity implies local stability in distribution economies, allow us to conclude that if the data is rationalizable in a Walrasian setting, then it is rationalizable in an economy in which each observation is locally stable under tâtonnement. Thus global instability, while clearly a theoretical possibility in Walrasian markets, cannot be detected in a finite data set consisting of observations on prices, incomes, and aggregate consumption.

The paper proceeds as follows. In Section 2 we discuss conditions for rationalizing individual demand in economies with smooth characteristics. By developing a set of "dual" Afriat inequalities, we show that if the observed data can be rationalized by an individual consumer with smooth characteristics then it can be rationalized by a smooth utility function which generates a locally monotone demand function. In Section 3 we discuss the implications of locally monotone demand and use these results together with the results from Section 2 to show that local uniqueness, local stability, and local monotone comparative statics are not refutable in Walrasian markets.

2 Rationalizing Individual Demand

Given a finite number of observations (p^r, x^r), $r = 1, \ldots, N$, on prices and quantities, when is this data consistent with utility maximization by some consumer with a nonsatiated utility function? We say that a utility function $U : \mathbb{R}^{\ell}_{+} \to \mathbb{R}$ *rationalizes* the data (p^r, x^r), $r = 1, \ldots, N$, if $\forall r$

$$p^r \cdot x^r \geq p^r \cdot x \Rightarrow U(x^r) \geq U(x), \ \forall x \in \mathbb{R}^{\ell}_{+}.$$

Using this terminology, we can restate the question above: given a finite data set, when does there exist a nonsatiated utility function which rationalizes these observations? The classic answer to this question was given by Afriat [Afr67].

Theorem (Afriat). *The following are equivalent:*

(a) *There exists a nonsatiated utility function which rationalizes the data.*
(b) *The data satisfies Cyclical Consistency.*
(c) *There exist numbers U^i, $\lambda^i > 0$, $i = 1, ..., N$ which satisfy the "Afriat inequalities":*

$$U^i - U^j \leq \lambda^j p^j \cdot (x^i - x^j) \quad (i, j = 1, ..., N).$$

(d) *There exists a concave, monotone, continuous, nonsatiated utility function which rationalizes the data.*

In particular, the equivalence of (a) and (d) shows that the hypothesis that preferences are represented by a concave utility function can never be refuted based on a finite data set, since if the data is rationalizable by any nonsatiated utility function then it is rationalizable by a concave, monotone, and continuous one. Moreover, Afriat showed explicitly how to construct such a utility function which rationalizes a given data set. For each $x \in \mathbb{R}_+^\ell$, define

$$U(x) = \min\{U^r + \lambda^r p^r \cdot (x - x^r)\}.$$

This utility function is indeed continuous, monotone and concave, and rationalizes the data by construction.

As Chiappori and Rochet [CR87] note, however, this utility function is piecewise linear and thus neither differentiable nor strictly concave. Such a utility function does not generate a smooth demand function, and for a number of prices does not even generate single-valued demand. This utility function is thus incompatible with many standard demand-based approaches to the question of rationalizability or estimation, as well as with our questions about comparative statics and asymptotic stability.

Whether or not a given set of observations can be rationalized by a smooth utility function which generates a smooth demand function will obviously depend on the nature of the observed data. Two situations in which such a rationalization is impossible are obvious: if two different consumption bundles are observed with the same price vector, or if one consumption bundle is observed with two different price vectors. If the data satisfies SARP, then this first case is eliminated; Chiappori and Rochet [CR87] show that if in addition this second case is ruled out then the data can be rationalized by a smooth, strongly concave utility function.

More formally, the data satisfies the *Strong Strong Axiom of Revealed Preference (SSARP)* if it satisfies SARP and if for all $i, j = 1, \ldots, N$,

$$p^i \neq p^j \Rightarrow x^i \neq x^j.$$

Chiappori and Rochet [CR87] show that given a finite set of data satisfying SSARP, there exists a strictly increasing, C^∞, strongly concave utility function defined on a compact subset of $\mathbb{R}^\ell_+$ which rationalizes this data. Although SSARP is a condition on the observable data alone, it can be equivalently characterized by the "strict Afriat inequalities." That is, the data satisfy SSARP if and only if there exist numbers $U^i, \lambda^i > 0$, $i = 1, \ldots, N$ such that

$$U^i - U^j < \lambda^j p^j \cdot (x^i - x^j) \quad (i, j = 1, \ldots, N, \ i \neq j).$$

Our work makes use of a modification of the results of Chiappori and Rochet [CR87]. Since our data consists of prices and income levels, we find it more natural to first recast the question of rationalizability in terms of indirect utility, and develop a set of dual Afriat inequalities characterizing data which can be rationalized by a consumer with smooth characteristics. An important benefit of this dual characterization is that it allows us to conclude not only that the data can be rationalized by a smooth demand function but by a demand function which is locally monotone in a neighborhood of each observation (p^r, I^r).

Definition. An individual demand function $f(p, I)$ is *locally monotone* at $(\bar{p}, \bar{I})$ if there exists a neighborhood $\mathcal{W}$ of $(\bar{p}, \bar{I})$ such that

$$(p - q) \cdot (f(p, I) - f(q, I)) < 0$$

for all $(p, I), (q, I) \in \mathcal{W}$ such that $p \neq q$.

Our first result can then be stated as follows.

Theorem 1. *Let (p^r, x^r), $r = 1, \ldots, N$ be given. There exists a utility function rationalizing this data that is strictly quasiconcave, monotone and smooth on an open set $\mathcal{X}$ containing x^r for $r = 1, \ldots, N$ such that the implied demand function is locally monotone at (p^r, I^r) for each $r = 1, \ldots, N$ where $I^r = p^r \cdot x^r$ if and only if there exist numbers V^i, λ^i, and vectors $q^i \in \mathbb{R}^\ell$, $i = 1, \ldots, N$ such that:*

(a) *for $i \neq j$,*

$$V^i - V^j > q^j \cdot \left(\frac{p^i}{I^i} - \frac{p^j}{I^j} \right) \quad (i, j = 1, \ldots, N)$$

(b) $\lambda^j > 0$, $q^j \ll 0$, $j = 1, \ldots, N$
(c) $q^j / I^j = -\lambda^j x^j$, $j = 1, \ldots, N$.

Conditions (a) and (b) constitute our "dual strict Afriat inequalities." Condition (c) here is just an expression of Roy's identity in this context. To see this, note that if (c) holds for some $\lambda^j > 0$, then $p^j \cdot q^j / I^j = -\lambda^j (p^j \cdot x^j) =$

$-\lambda^j I^j$, i.e., $\lambda^j = -p^j \cdot q^j/(I^j)^2$, which implies that the vector $(q^j/I^j, \lambda^j)$ corresponds to the gradient of the rationalizing indirect utility function V evaluated at (p^j, I^j). This is essentially the content of (a). More precisely, (a) says essentially that q^j is the derivative of V with respect to the normalized price vector p/I evaluated at (p^j, I^j). Thus

$$\frac{\partial V}{\partial p}(p^j, I^j) = \frac{q^j}{I^j} \ \text{ and } \ \frac{\partial V}{\partial I}(p^j, I^j) = -\frac{p^j \cdot q^j}{(I^j)^2}.$$

Then if $q^j/I^j = -\lambda^j x^j$, x^j is indeed demand at the price-income pair (p^j, I^j) by Roy's identity.

The proof of Theorem 1 relies on two intermediate results. The first is a version of Lemma 2 in Chiappori and Rochet [CR87] modified to apply to our dual Afriat inequalities.

Lemma 1. *If there exist numbers V^i, λ^i and vectors q^i, $i = 1, \ldots, N$ satisfying the dual strict Afriat inequalities, then there exists a convex, homogeneous of degree 0, C^∞ function $W : \mathbb{R}_+^{\ell+1} \to \mathbb{R}$ which is strictly increasing in I and strictly decreasing in p such that $W(p^i, I^i) = V^i$, $DW(p^i, I^i) = (q^i/I^i, \lambda^i)$, and $D_{pp}^2 W(p^i, I^i) = 0$ for every $i = 1, \ldots, N$.*

Proof. For each $(p, I) \in \mathbb{R}_{++}^{\ell+1}$ define

$$Y(p, I) = \max_t \left\{ V^t + q^t \cdot \left(\frac{p}{I} - \frac{p^t}{I^t} \right) \right\}.$$

Then Y is convex in p, homogeneous of degree 0 in (p, I), continuous, strictly increasing in I and strictly decreasing in p. Moreover, $Y(p^r, I^r) = V^r$ for each r. To see this, note that by definition $\forall r \ \exists s$ such that

$$Y(p^r, I^r) = V^s + q^s \cdot \left(\frac{p^r}{I^r} - \frac{p^s}{I^s} \right).$$

By the dual strict Afriat inequalities, if $r \neq s$ then

$$V^r - V^s > q^s \cdot \left(\frac{p^r}{I^r} - \frac{p^s}{I^s} \right).$$

So

$$\forall s \neq r, V^r > V^s + q^s \cdot \left(\frac{p^r}{I^r} - \frac{p^s}{I^s} \right).$$

Thus $Y(p^r, I^r) = V^r$ for each $r = 1, \ldots, N$.

This argument shows that for every $s \neq r$,

$$Y(p^r, I^r) > V^s + q^s \cdot \left(\frac{p^r}{I^r} - \frac{p^s}{I^s} \right).$$

Since Y is continuous, $\forall r$ there exists $\eta_r > 0$ such that

$$Y(p, I) = V^r + q^r \cdot \left(\frac{p}{I} - \frac{p^r}{I^r} \right)$$

for all $(p, I) \in B((p^r, I^r), \eta_r)$, i.e., Y is piecewise linear in p/I, and hence in p. Following Chiappori and Rochet [CR87], we smooth Y by convolution, which yields a function W which is C^∞, homogeneous of degree 0, convex in p/I and hence convex in p, strictly decreasing in p and strictly increasing in I. Moreover, $D_{pp}^2 W(p^r, I^r) = 0$ for $r = 1, \ldots, N$. □

The intuition behind this result is straightforward. The dual Afriat function $Y(p, I)$ gives an indirect utility function rationalizing the data that is locally linear in p/I and that has kinks whenever two or more of the dual Afriat equations are equal. Since the data satisfy the strict dual Afriat inequalities, none of these kinks occurs at an observation, so smoothing the function Y in a sufficiently small neighborhood of each kink gives a smooth function which is equal to the original dual Afriat function in a neighborhood of each observation, and thus in particular is locally linear in prices in each such neighborhood.

The second important result we use, due to Quah [Qua98], gives conditions on indirect utility analogous to the Mijutschin–Polterovich conditions on direct utility under which individual demand is locally monotone.

Theorem (Quah, Theorem 2.2). *Let $V : \mathbb{R}_+^{\ell+1} \to \mathbb{R}$ be a smooth, strictly quasiconvex indirect utility function that is convex in p and satisfies the Elasticity Condition[4]*

$$-\frac{p^T D_{pp}^2 V(\bar{p}, \bar{I}) p}{p^T D_p V(\bar{p}, \bar{I})} < 4$$

or equivalently,

$$e(\bar{p}, \bar{I}) \equiv \frac{\bar{I} V_{II}(\bar{p}, \bar{I})}{V_I(\bar{p}, \bar{I})} < 2.$$

Then V generates a demand function which is locally monotone in a neighborhood of $(\bar{p}, \bar{I})$.

In particular, if the indirect utility function V is linear in prices in a neighborhood of $(\bar{p}, \bar{I})$, then the Elasticity Condition is clearly satisfied, and thus Quah's theorem shows that if the indirect utility function is "nearly" locally linear in prices then it generates a demand function that is locally monotone in a neighborhood of $(\bar{p}, \bar{I})$.

The proof of Theorem 1 then combines these two ideas and exploits the duality between direct and indirect utility.

Proof of Theorem 1. First suppose there exist solutions to the inequalities (a), (b) and (c). Let

---

[4] Here subscripts denote partial derivatives.

$$M = \max_{s,t} \left\| \frac{p^s}{I^s} - \frac{p^t}{I^t} \right\|^2.$$

Since there exists a solution to the strict dual Afriat inequalities, there exists $\varepsilon > 0$ sufficiently small so that $\forall i \neq j$,

$$V^i - V^j > q^j \cdot \left(\frac{p^i}{I^i} - \frac{p^j}{I^j} \right) + \varepsilon M. \tag{$*$}$$

Define

$$\tilde{q}^j = q^j - \varepsilon \frac{p^j}{I^j} \quad (j = 1, \ldots, N)$$

and

$$\tilde{V}^j = V^j - \tfrac{1}{2}\varepsilon \left\| \frac{p^j}{I^j} \right\|^2 \quad (j = 1, \ldots, N).$$

Then by (2),[5] $\forall i \neq j$

$$\tilde{V}^i - \tilde{V}^j > \tilde{q}^j \cdot \left(\frac{p^i}{I^i} - \frac{p^j}{I^j} \right).$$

Now by Lemma 1, there exists a C^∞ function $W : \mathbb{R}_+^{\ell+1} \to \mathbb{R}$ that is homogeneous of degree 0, convex in p/I, strictly increasing in I, strictly decreasing in p and satisfies:

$$W(p^i, I^i) = \tilde{V}^i \quad (i = 1, \ldots, N)$$
$$D_{p/I} W(p^i, I^i) = \tilde{q}^i \quad (i = 1, \ldots, N)$$
$$D_{pp}^2 W(p^i, I^i) = 0 \quad (i = 1, \ldots, N).$$

Define $V(p, I) = W(p, I) + \tfrac{1}{2}\varepsilon \| p/I \|^2$. Let $\mathcal{P}$ be a compact, convex subset of $\mathbb{R}_+^\ell$ containing 0 and containing p^i/I^i in its interior for $i = 1, \ldots, N$, and let $\hat{\mathcal{P}} \equiv \{(p, I) : p/I \in \mathcal{P}\}$. Then V is C^∞, homogeneous of degree 0, strictly convex in p/I and hence in p, strictly increasing in I and strictly decreasing in p on $\hat{\mathcal{P}}$ for ε sufficiently small. Moreover,

$$V(p^i, I^i) = V^i \quad (i = 1, \ldots, N_-$$
$$D_{p/I} V(p^i, I^i) = q^i \quad (i = 1, \ldots, N)$$
$$D_{pp}^2 V(p^i, I^i) = \frac{\varepsilon}{(I^i)^2} id \quad (i = 1, \ldots, N),$$

where id is the $\ell \times \ell$ identity matrix. For ε sufficiently small, $e(p^i, I^i) < 2$ for all $i = 1, \ldots, N$. By Quah's theorem, the demand function

$$x(p, I) \equiv -\frac{1}{D_I V(p, I)} D_p V(p, I)$$

---

[5] See Chiappori and Rochet [CR87] for the details.

generated by this indirect utility function is locally monotone at (p^i, I^i) for each $i = 1, \ldots, N$.

To convert this indirect utility function into a direct utility function, for each $x \in \mathbb{R}_+^\ell$ define

$$U(x) \equiv \min_{p \in \mathcal{P}} V(p, 1) \quad \text{subject to} \quad p \cdot x \leq 1. \tag{$\dagger$}$$

Let $p(x)$ denote the solution to (2) given x, and $\mathcal{X} \equiv p^{-1}(\mathcal{P}^o)$. Then by standard arguments, U is strictly quasiconcave, monotone, and smooth on $\mathcal{X}$. Moreover, $x(p, I)$ is the demand function generated by U on $\mathcal{P}^o$, so U rationalizes the data. To establish this claim, we must show that for each $p \in \mathcal{P}^o$,

$$V(p, 1) = \max_{x \in \mathbb{R}_+^\ell} U(x) \quad \text{subject to} \quad p \cdot x \leq 1.$$

Note that by the definition of U, for any $p \in \mathcal{P}$ and $x > 0$ such that $p \cdot x \leq 1$, $V(p, 1) \geq U(x)$, so

$$V(p, 1) \geq \max_{x \in \mathbb{R}_+^\ell} U(x).$$

Then it suffices to show that given $\bar{p} \in \mathcal{P}^o$ there exists $\bar{x} > 0$ such that $\bar{p} \cdot \bar{x} \leq 1$ and $V(\bar{p}, 1) = U(\bar{x})$. Let $C = \{p \in \mathcal{P} : V(p, 1) \leq V(\bar{p}, 1)\}$. Since V is continuous and quasiconvex, C is closed and convex. Since V is strictly decreasing in p, $\bar{p}$ is a boundary point of C. By the supporting hyperplane theorem, $\exists \bar{x} \neq 0$ such that $\bar{p} \cdot \bar{x} \leq p \cdot \bar{x}$ for all $p \in C$. Moreover, $\bar{x} > 0$ since $\forall i \ \exists \alpha_i > 0$ such that $\bar{p} + \alpha_i b_i \in C$, where b_i is the ith unit vector, and, by rescaling $\bar{x}$ if necessary, $\bar{p} \cdot \bar{x} = 1$. Then $\bar{p}$ solves ($\dagger$) given $\bar{x}$, since if not, then there exists $p \in \mathcal{P}$ such that $p \cdot \bar{x} \leq \bar{p} \cdot \bar{x} = 1$ and $V(p, 1) < V(\bar{p}, 1)$. Then $p \cdot \bar{x} = \bar{p} \cdot \bar{x} = 1$, so there exists $\alpha \in (0, 1)$ such that $(\alpha p) \cdot \bar{x} < \bar{p} \cdot \bar{x}$ and $V(\alpha p, 1) < V(\bar{p}, 1)$. Since $\mathcal{P}$ is convex and contains 0, $\alpha p \in \mathcal{P}$. This contradicts the definition of $\bar{x}$, however. Thus $V(\bar{p}, 1) = U(\bar{x})$. Finally, by Roy's identity $x(p, I)$ is the demand function generated by U on $\mathcal{P}^o$, which also implies that U rationalizes the data.

Now suppose there exists a smooth, strictly quasiconcave, monotone utility function rationalizing $(p^r, x^r), r = 1, \ldots, N$. Let V be the indirect utility function generated by this utility function. Then by the standard duality arguments given above, V rationalizes $(p^r/I^r, x^r), r = 1, \ldots, N$ where $I^r = p^r \cdot x^r$ for each r, in the sense that if $-V(p, I) > -V(p^r, I^r)$ then $p \cdot x^r > I$. Equivalently, if $-V(p/I, 1) > -V(p^r/I^r, 1)$ then $x^r \cdot (p/I) > 1 = x^r \cdot (p^r/I^r)$ for $r = 1, \ldots, N$. Then applying Chiappori and Rochet's result to $-V$ and interchanging the roles of p and x implies that there exist numbers $V^j, \gamma^j > 0$, $j = 1, \ldots, N$ such that for all $i \neq j$,

$$V^i - V^j > -\gamma^j x^j \cdot \left(\frac{p^i}{I^i} - \frac{p^j}{I^j} \right) \quad (i, j = 1, \ldots, N).$$

Define $\lambda^j = \gamma^j/I^j$, $j = 1, \ldots, N$. Then substituting $q^j \equiv -\gamma^j x^j = -\lambda^j I^j x^j$, $j = 1, \ldots, N$, gives the desired inequalities. $\qquad \square$

The conditions that are necessary and sufficient for the existence of a smooth, monotone utility function rationalizing a given finite data set—the dual strict Afriat inequalities—are exactly the same conditions that are necessary and sufficient for the existence of smooth preferences rationalizing the data for which demand is locally monotone at each observation. Thus it follows immediately from Theorem 1 that any finite data set that can be rationalized by smooth preferences can be rationalized by smooth preferences giving rise to locally monotone demand.

Corollary. *Let (p^r, x^r), $r = 1, \ldots, N$ be given. There exists a smooth, strictly quasiconcave, monotone utility function rationalizing this data if and only if there exists a smooth, strictly quasiconcave, monotone utility function rationalizing this data such that the implied demand function is locally monotone at (p^r, I^r) for each $r = 1, \ldots, N$, where $I^r = p^r \cdot x^r$.*

This result will provide the foundation for the study of rationalizing equilibria which is contained in the next section.

3 Rationalizing Walrasian Equilibria

In this section, we turn to the question of rationalizing observations as equilibria. Here we consider a finite number of observations $\langle p^r, \omega^r, \{I_t^r\}_{t=1}^T \rangle$, $r = 1, \ldots, N$ of prices, aggregate consumption, and income levels for each of T consumers. Our main result shows that such a finite data set can never be used to refute the hypotheses that equilibria are locally unique or locally stable under tâtonnement, or that equilibrium comparative statics are locally monotone.

Theorem 2. *Let $\langle p^r, \omega^r, \{I_t^r\}_{t=1}^T \rangle$, $r = 1, \ldots, N$, be given. This data can be rationalized by an economy in which each consumer has a smooth, strictly quasiconcave, monotone utility function if and only if it can be rationalized by an economy in which each consumer has a smooth, strictly quasiconcave, monotone utility function and in which each observed equilibrium p^r is locally unique and locally stable under tâtonnement and in which the equilibrium correspondence is locally monotone at (p^r, ω^r) for each r.*

To establish this result, consider a given finite data set $\langle p^r, \omega^r, \{I_t^r\}_{t=1}^T \rangle$, $r = 1, \ldots, N$. When can these observations be rationalized as Walrasian equilibria? Since we do not observe individual consumption bundles or utilities, these observations are *rationalizable as Walrasian equilibria* if for each observation $r = 1, \ldots, N$ there exist consumption bundles x_t^r for each consumer $t = 1, \ldots, T$ such that the individual observations (p^r, x_t^r), $r = 1, \ldots, N$, are rationalizable for each consumer, $p^r \cdot x_t^r = I_t^r$ for each r and t, and such that markets clear in each observation, that is, $\sum_{t=1}^T x_t^r = \omega^r$ for each r. Putting this definition together with the dual strict Afriat inequalities characterizing individual rationalizability yields the following result.

Lemma 2. *Let* $\langle p^r, \omega^r, \{I_t^r\}_{t=1}^T \rangle$, $r = 1, \ldots, N$ *be a finite set of observations. There exist smooth, strictly quasiconcave, monotone utility functions rationalizing this data and initial endowments* $\{\omega_t^r\}_{t=1}^T$ *such that* p^r *is an equilibrium price vector for the economy* E^r *if and only if there exist numbers* V_t^r, λ_t^r *and vectors* q_t^r *for* $t = 1, \ldots, T$ *and* $r = 1, \ldots, N$ *such that:*

(a) *the dual strict Afriat inequalities hold for each consumer* $t = 1, \ldots, T$,
(b) $p^r \cdot q_t^r = -\lambda_t^r (I_t^r)^2$ *for* $t = 1, \ldots, T$ *and* $r = 1, \ldots, N$
(c) *"markets clear":*

$$\sum_{t=1}^T x_t^r = \omega^r \quad \forall r$$

where $x_t^r = -(1/\lambda_t^r)(q_t^r/I_t^r)$ *for each* r *and* t.

Proof. Follows immediately from Theorem 1 and the definition of rationalizability. □

Moreover, note that given a finite set of observations $\langle p^r, \omega^r, \{I_t^r\}_{t=1}^T \rangle$, $r = 1, \ldots, N$, without observations of initial endowments we can without loss of generality assume that each observation of the income distribution $\{I_t^r\}_{t=1}^T$ is derived from collinear individual endowments. More precisely, for each r and t define

$$\alpha_t^r = \frac{I_t^r}{p^r \cdot \omega^r}$$

and $\alpha^r = (\alpha_1^r, \ldots, \alpha_T^r)$. Given utility functions $\{U_t\}_{t=1}^T$, the *distribution economy* $\mathcal{E}_{\alpha^r}$ is the economy in which consumer t has preferences represented by the utility function U_t and endowment $\alpha_t^r \omega^r$. Using this observation we can now restate Lemma 2 as follows: given a finite set of observations $\langle p^r, \omega^r, \{I_t^r\}_{t=1}^T \rangle$, $r = 1, \ldots, N$, there exist smooth, strictly quasiconcave, monotone utility functions rationalizing this data such that p^r is an equilibrium price vector for the distribution economy $\mathcal{E}_{\alpha^r}$ if and only if there exist numbers V_t^r, λ_t^r and vectors q_t^r for $t = 1, \ldots, T$ and $r = 1, \ldots, N$ such that conditions (a), (b), and (c) of Lemma 2 hold.

The dual strict Afriat inequalities in (a) are exactly the conditions characterizing observations which can be rationalized by consumers with smooth characteristics, and we showed in Theorem 1 that such observations can always be rationalized by utility functions generating locally monotone demand. The results we derive regarding the refutability of local stability and local comparative statics then follow from the striking properties of locally monotone individual demand functions.

First, unlike almost every other property of individual demand such as the weak axiom or the Slutsky equation, local monotonicity aggregates. If f_t is an individual demand function which is locally monotone at $(\bar{p}, \bar{I}_t)$ for each $t = 1, \ldots, T$, then market excess demand

$$F(p) \equiv \sum_{t=1}^{T} f_t(p, \bar{I}_t) - \omega$$

is locally monotone at $\bar{p}$.

Furthermore, local monotonicity at equilibrium implies local uniqueness[6] and, in distribution economies, local stability of tâtonnement and local monotone comparative statics, as the following result shows.

Theorem (Malinvaud). *Let $\bar{p}$ be an equilibrium price vector for a distribution economy $\mathcal{E}_\alpha$ with income distribution $\{\alpha_t\}_{t=1}^{T}$. If each consumer's demand function is locally monotone at $(\bar{p}, \bar{I}_t)$, where $\bar{I}_t = \alpha_t(\bar{p} \cdot w)$, then the tâtonnement price adjustment process is locally stable at $\bar{p}$. Furthermore, the equilibrium correspondence $\mathcal{C}$ for the distribution economy $\mathcal{E}_\alpha$[7] is locally monotone in a neighborhood $\mathcal{P} \times \mathcal{U}$ of $(\bar{p}, w)$ i.e., if $(p', w') \in (\mathcal{P} \times \mathcal{U}) \cap \mathcal{C}$ then*

$$(p' - p) \cdot (w' - w) < 0.$$

This conclusion of locally monotone comparative statics implies in particular that if the aggregate supply of a good increases, all else held constant, then its equilibrium price will fall, at least locally.[8]

The main conclusion of the paper is now an immediate consequence of the results of Section 2 and the properties of locally monotone demand functions in distribution economies: local uniqueness, local stability, and local monotone comparative statics are not refutable given a finite set of observations of prices, income levels, and aggregate consumption.[9]

Proof of Theorem 2. Let $\langle p^r, w^r, \{I_t^r\}_{t=1}^{T} \rangle$, $r = 1, \ldots, N$, be a finite set of observations which can be rationalized in an economy in which each consumer has a smooth, strictly quasiconcave, monotone utility function. Then conditions (a), (b), and (c) of Lemma 2 must hold. By Theorem 1, there exist smooth, strictly quasiconcave, monotone utility functions rationalizing the data such that p^r is an equilibrium price vector for the distribution economy $\mathcal{E}_{\alpha^r}$ and such that the market excess demand function for $\mathcal{E}_{\alpha^r}$ is locally monotone at p^r for each $r = 1, \ldots, N$. Thus p^r is a locally unique equilibrium in the

---

[6] To see this, note that if $F(\bar{p}) = 0$ and F is locally monotone at $\bar{p}$, then $(p - \bar{p}) \cdot F(p) < 0$ for p sufficiently close to $\bar{p}$. In particular, $F(p) \neq 0$ for such p.

[7] Here the income distribution is assumed to be constant as aggregate endowment changes, so the equilibrium correspondence C for a distribution economy E_α is the set of pairs (p, w) such that p is an equilibrium price for the economy in which consumer t has utility U_t and endowment $\alpha_t w$.

[8] For this result, prices are normalized so that the value of the aggregate endowment is constant, which means the prices p and p^r are subject to different normalizations. As Nachbar [Nac02] points out, this makes the interpretation of this comparative statics result problematic. See Nachbar [Nac02] for further discussion.

[9] Note that this result is still valid if in addition individual demands are observed.

economy $\mathcal{E}_{\alpha^r}$. By Malinvaud's results, p^r is locally stable under tâtonnement, and the equilibrium correspondence in the distribution economy $\mathcal{E}_{\alpha^r}$ is locally monotone at (p^r, ω^r) for each $r = 1, \ldots, N$.   □

Acknowledgments

We are pleased to acknowledge the insightful comments of John Quah and two referees on previous versions. Shannon's work was supported in part by NSF Grant SBR-9321022 and by an Alfred P. Sloan Foundation Research Fellowship. This work was begun while Shannon was on leave at Yale University; the warm hospitality of the Economics Department and the Cowles Foundation is greatly appreciated.

Brown, D.J., Shannon, C.: Uniqueness,stability and comparative statics in rationalizable Walrasian markets. Econmetrica **64**, 1249–1262 (1996). Reprinted by permission of the Econometric Society.

The Nonparametric Approach to Applied Welfare Analysis

Donald J. Brown[1] and Caterina Calsamiglia[2]

[1] Yale University, New Haven, CT 06511 donald.brown@yale.edu
[2] Universitat Autònoma de Barcelona, Bellaterra, Barcelona, Spain 08193
caterina.calsamiglia@uab.es

Summary. Changes in total surplus are traditional measures of economic welfare. We propose necessary and sufficient conditions for rationalizing individual and aggregate consumer demand data with individual quasilinear and homothetic utility functions. Under these conditions, consumer surplus is a valid measure of consumer welfare. For nonmarketed goods, we propose necessary and sufficient conditions on input market data for efficient production , i.e. production at minimum cost. Under these conditions we derive a cost function for the nonmarketed good, where producer surplus is the area above the marginal cost curve.

Key words: Welfare economics, Quasilinear utilities, Homothetic utilities, Non-marketed goods, Afriat inequalities

1 Introduction

Theoretical models of consumer demand assume consumer's choice is derived from the individual maximizing a concave utility function subject to a budget constraint. Given a finite set of price and consumption choices, we say that the data set is rationalizable if there exists a concave, continuous and monotonic utility function such that each choice is the maximum of this function over the budget set.

Afriat [Afr67] provided the first necessary and sufficient conditions for a finite data set to be a result of the consumer maximizing her utility subject to a budget constraint. In a series of lucid papers Varian [Var82, Var83, Var84] increased our understanding and appreciation of Afriat's seminal contributions to demand theory, Afriat [Afr67], and the theory of production, Afriat [Afr72a]. As a consequence there is now a growing literature on the testable restrictions of strategic and non-strategic behavior of households and firms in market economies—see the survey of Carvajal, et al. [CRS04].

Quasilinear utility functions are used in a wide range of areas in economics, including theoretical mechanism design, public economics, industrial organization and international trade. In applied partial equilibrium models we often measure economic welfare in terms of total surplus, i.e., the sum of consumer and producer surplus, and deadweight loss. As is well known, consumer surplus is a valid measure of consumer welfare only if consumer's demand derives from maximizing a homothetic or quasilinear utility function subject to her budget constraint—see section 11.5 in Silberberg [Sil90]. Both Afriat [Afr72b] and Varian [Var83] proposed a necessary and sufficient combinatorial condition for rationalizing data sets, consisting of market prices and consumer demands, with homothetic utility functions. This condition is the homothetic axiom of revealed preference or HARP. To our knowledge, there is no comparable result in the literature for quasilinear rationalizations of consumer demand data. In this paper we show that a combinatorial condition introduced in Rockafellar [Roc70] to characterize the subgradient correspondence for convex real-valued functions on $\mathbb{R}^n$, i.e., cyclical monotonicity, is a necessary and sufficient condition for a finite data set to be rationalizable with a quasilinear utility function.[3] We then extend our analysis to aggregate demand data. That is, given aggregate demands, the distribution of expenditures of consumers and market prices we give necessary and sufficient conditions for the data to be rationalized as the sum of individual demands derived from maximizing quasilinear utility functions subject to a budget constraint. Our analysis differs from Varian's in that it is the form of the Afriat inequalities for the homothetic and quasilinear case that constitute the core of our analysis. We show in the case of aggregate data, where individual demand data is not observed, that the Afriat inequalities for the homothetic case reduce to a family of convex inequalities and for the quasilinear case they reduce to linear inequalities. As a consequence in both cases we can determine in polynomial time if they are solvable and, if so, compute the solution in polynomial time. This is certainly not true in general.[4]

On the other hand, measuring producer surplus for nonmarketed goods such as health, education or environmental amenities and ascertaining if these goods are produced efficiently, i.e., at minimum cost, are important policy issues. Our contribution to the literature on nonmarketed goods is the observation that Afriat's combinatorial condition, cyclical consistency or CC, and equivalently Varian's generalized axiom of revealed preference or GARP, are necessary and sufficient conditions for rationalizing a finite data set, consisting of factor demands and factor prices, with a concave, monotone and continuous production function. Hence they constitute necessary and sufficient conditions for nonmarketed goods to be produced at minimum cost for some production function. If these conditions hold, then the supply curve for the nonmarketed

---

[3] In the paper by Rochet [Roc87] "A necessary and sufficient condition for rationalizability in a quasilinear context" published in the *Journal of Mathematical Economics* he defines rationalizability as implementability of an action profile via compensatory transfers. Therefore the results presented in his paper are of a different nature.

[4] See Brown and Kannan [BK06] for further discussion.

good is the marginal cost curve of the associated cost function and producer surplus is well defined.

2 Rationalizing Individual Demand Data with Quasilinear Utilities

In this section we provide necessary and sufficient conditions for a data set to be the result of the consumer maximizing a quasilinear utility function subject to a budget constraint.

Definition 1. Let (p_r, x_r), $r = 1, ..., N$ be given. The data is *quasilinear rationalizable* if for some $y_r > 0$ and $I > 0$, $\forall r$ x_r solves

$$\max_{x \in \mathbb{R}^n_{++}} U(x) + y_r \quad \text{s.t.} \quad p_r x + y_r = I.$$

for some concave U and numeraire good y_r.

Theorem 1. *The following are equivalent:*

(1) *The data (p_r, x_r), $r = 1, ..., N$ is quasilinear rationalizable by a continuous, concave, strictly monotone utility function U.*

(2) *The data (p_r, x_r), $r = 1, ..., N$ satisfies Afriat's inequalities with constant marginal utilities of income, that is, there exists $G_r > 0$ and $\lambda > 0$ for $r = 1, ..., N$ such that*

$$G_r \leq G_\ell + \lambda p_\ell \cdot (x_r - x_\ell) \quad \forall r, \ell = 1, ..., N$$

or equivalently there exist $U_r > 0$ for $r = 1, ..., N$

$$U_r \leq U_\ell + p_\ell \cdot (x_r - x_\ell) \quad \forall r, \ell = 1, ..., N$$

where $U_r = G_r / \lambda$.

(3) *The data (p_r, x_r), $r = 1, ..., N$ is "cyclically monotone," that is, if for any given subset of the data $\{(p_s, x_s)\}_{s=1}^m$:*

$$p_1 \cdot (x_2 - x_1) + p_2 \cdot (x_3 - x_2) + \cdots + p_m \cdot (x_1 - x_m) \geq 0.$$

Proof.
(1) $\Rightarrow$ (2): First note that if U is concave on $\mathbb{R}^n$, then $\beta \in \mathbb{R}^n$ is a subgradient of U at x if for all $y \in \mathbb{R}^n : U(y) \leq U(x) + \beta \cdot (y - x)$. Let $\partial U(x)$ be the set of subgradients of U at x. From the FOC of the quasilinear utility maximization problem we know:

$$\exists \beta_r \in \partial U(x) \quad \text{s.t.} \quad \beta_r = \lambda_r p_r \text{ where } \lambda_r = 1.$$

Also, U being concave implies that $U(x_r) \leq U(x_\ell) + \beta_\ell(x_r - x_\ell)$ for $r, \ell = 1, 2, \ldots, N$. Since $\beta_\ell = p_\ell \ \forall \ell = 1, \ldots, N$ we get $U(x_r) \leq U(x_\ell) + p_\ell(x_r - x_\ell) \ \forall r, \ell = 1, \ldots, N$.

(2) $\Rightarrow$ (3): For any set of pairs $\{(x_s, p_s)\}_{s=1}^m$ we need that: $p_0 \cdot (x_1 - x_0) + p_1(x_2 - x_1) + \cdots + p_m(x_0 - x_m) \geq 0$.

From the Afriat inequalities with constant marginal utilities of income we know:

$$U_1 - U_0 \ \leq p_0 \cdot (x_1 - x_0)$$

$$\vdots$$

$$U_0 - U_m \leq p_m \cdot (x_0 - x_m).$$

Adding up these inequalities we see that the left hand sides cancel and the resulting condition defines cyclical monotonicity.

(3) $\Rightarrow$ (1): Let $U(x) = \inf\{p_m \cdot (x - x_m) + \cdots + p_1 \cdot (x_2 - x_1)\}$ where the infimum is taken over all finite subsets of data, then $U(x)$ is a concave function on $\mathbb{R}^n$ and p_r is the subgradient of U at $x = x_r$ (this construction is due to Rockafellar [Roc70] in his proof of Theorem 24.8). Hence if $\lambda_r = 1$ for $r = 1, \ldots, N$ then $p_r = \partial U(x_r)$ constitutes a solution to the first order conditions of the quasilinear maximization problem.

If we require strict inequalities in (2) of Theorem 1, then it follows from Lemma 2 in Chiappori and Rochet [CR87] that the rationalization can be chosen to be a C^∞ function. It then follows from Roy's identity that $x(p) = -\partial V(p)/\partial p$. Hence for any line integral we see that $\int_{p_1}^{p_2} x(p)dp = -\int_{p_1}^{p_2} [\partial V(p)/\partial p]dp = V(p_1) - V(p_2)$. That is, consumer welfare is well-defined and the change in consumer surplus induced by a change in market prices is the change in consumer's welfare. $\qquad \Box$

3 Rationalizing Aggregate Demand Data with Quasilinear and Homothetic Utilities

Consumer surplus is the area under the aggregate demand function. This function is estimated from a finite set of observations of market prices and aggregate demand. Assuming we have the distribution of expenditures of consumers for each observation, we can derive the Afriat inequalities for each household, where individual consumptions are unknown, but are required to sum up to the observed aggregate demand. Setting marginal utilities of income equal to one, we have a system of linear inequalities in utility levels and individual demands, which can be solved in polynomial time using interior-point methods. For the homothetic case, the Afriat inequalities reduce to $U_i \leq U_j p_j \cdot (x_i - x_j) \ \forall i, j$. Changing variables where $U_i = e^{z_i}$ we derive a family of smooth convex inequalities of the form $e^{z_i - z_j} - p_j \cdot (x_i - x_j) \leq 0$, which can also be solved in polynomial time using interior point methods.

4 Rationalizing the Production of Nonmarketed Goods

Health, education and environmental amenities are all examples of nonmarketed goods. To compute producer surplus for such goods, we must derive the supply curve, given only factor demand and prices, since demand data and prices for these goods are not observed. An important policy issue is whether these goods are produced efficiently given the factor demand and prices. In fact, as we show, there may be no concave, monotone and continuous production function that rationalizes the input data. If one does exist, we can rationalize the data and derive the supply curve for the nonmarketed good.[5]

Definition 2. Let (p_r, x_r), $r = 1, \ldots, N$ be given. A production, F, rationalizes the data if for all $r = 1, \ldots, N$ there exists q_r such that x_r solves:

$$\max_{x \in \mathbb{R}^n_{++}} q_r F(x) - p_r \cdot x$$

where F is a concave function.

Theorem 2. *The following conditions are equivalent:*

(1) *There exists a concave, monotone, continuous, non-satiated production function that rationalizes the data.*
(2) *The data (p_r, x_r), $r = 1, ..., N$ satisfies Afriat inequalities, that is, there exists $F_r > 0$ and $q_r > 0$ for $r = 1, \ldots, N$ such that*

$$F_r \leq F_\ell + \frac{1}{q_\ell} p_\ell \cdot (x_r - x_\ell) \ \ \forall r, \ell = 1, \ldots, N$$

where q_ℓ is the marginal cost of producing F_ℓ.
(3) *The data (p_r, x_r), $r = 1, ..., N$ satisfies "cyclical consistency," that is,*

$$p_r x_r \geq p_r x_s, p_s x_s \geq p_s x_t, ..., p_q x_q \geq p_q x_r$$

implies

$$p_r x_r = p_r x_s, p_s x_s = p_s x_t, ..., p_q x_q = p_q x_r.$$

Proof. This is Afriat's [Afr67] result where we let $F = U$ and $\lambda_r = 1/q_r$. $\square$

---

[5] If we write the cost minimization problem of the firm, $\min_{x \in \mathbb{R}^n} p \cdot x$ s.t. $F(x) \geq y$, from the F.O.C. we find $p = \mu F'(x)$, where μ is the Lagrange multiplier associated with the constraint, and therefore is equal to the marginal cost of producing one more unit of output at the optimum. Therefore it is easy to see from the FOC of the profit maximization problem that the output price, q_r, is the marginal cost of production. The inequalities in (2) in Theorem 2 are the same as those in condition (3) of Theorem 2 in Varian [Var84], where he assumes that the production levels F_r are observable.

$F(x) = \min_{1 \leq \ell \leq r} \{ F_\ell + (1/q_\ell) p_\ell (x - x_\ell) \}$ is Afriat's utility (production) function derived from a solution to the Afriat inequalities. The associated expenditure (cost) function is $c(y; p) = \min_{x \in \mathbb{R}^n} p \cdot x$ s.t. $F(x) \geq y$. In the production setting, the supply curve is the marginal cost curve.

Acknowledgments

The work is partially supported by the Spanish Ministry of Science and Technology, through Grant BEC2002-2130 and the Barcelona Economics Program (CREA).

Brown, D.J., Calsamiglia, C.: The nonparametric approach to applied welfare analysis. Economic Theory **31**, 183-188 (2007). Reprinted by permission of Springer-Verlag.

Competition, Consumer Welfare, and the Social Cost of Monopoly

Yoon-Ho Alex Lee[1] and Donald J. Brown[2]

[1] U.S. Securities & Exchange Commission, Washington, DC 20549
  alex.lee@aya.yale.edu
[2] Yale University, New Haven, CT 06511 donald.brown@yale.edu

Summary. Conventional deadweight loss measures of the social cost of monopoly ignore, among other things, the social cost of inducing competition and thus cannot accurately capture the loss in social welfare. In this Article, we suggest an alternative method of measuring the social cost of monopoly. Using elements of general equilibrium theory, we propose a social cost metric where the benchmark is the Pareto optimal state of the economy that uses the least amount of resources, consistent with consumers' utility levels in the monopolized state. If the primary goal of antitrust policy is the enhancement of consumer welfare, then the proper benchmark is Pareto optimality, not simply competitive markets. We discuss the implications of our approach for antitrust law as well as how our methodology can be used in practice for allegations of monopoly power given a history of price-demand observations.

Key words: Monopoly power, Antitrust economics, Applied general equilibrium

1 Introduction

Monopoly and market power constitute the backbone of federal antitrust law. The Sherman Act[3]—largely regarded as the origin of the federal antitrust law and passed in 1890—was the government's response to cartelization and monopolization. Section 2 of the Sherman Act specifically prohibits monopolization as well as attempts to monopolize. In modern antitrust law, the existence of monopoly power is one of the two essential elements of the *Grinnell* test,[4] a test that is applied in all Section 2 cases of the Sherman Act.[5] Proof of market power is also required for antitrust violation under Section 7 of the Clayton Act.[6] Judge Richard A. Posner [Pos01, p. 9] argues that

---

[3] 15 U.S.C. §2 (1976).

[4] See United States v. Grinnell Corp., 384 U.S. 563 (1966).

[5] For a brief description of the development of Section 2 of the Sherman Act, see Hylton [Hyl03].

[6] 15 U.S.C. §18 (1976).

"the economic theory of monopoly provides the only sound basis for antitrust policy". That antitrust scholars are mindful of the social cost of monopoly and market power is also illustrated, for instance, by Professor William M. Landes and Judge Posner's remark [LP81] that the *size of the market* should be a determinant factor in judging whether a certain degree of market power should be actionable under antitrust law (see [LP81]). They note that "the actual economic injury caused to society is a function of [the size of the market]" and "[i]f the amount of economic activity is small, the total social loss is small, and an antitrust proceeding is unlikely to be socially cost justified." Accordingly, a clear understanding and a workable definition of the social cost of monopoly are essential in shaping and implementing antitrust law.

A familiar measure of the social cost of monopoly is the deadweight loss triangle—the social surplus unrealized due to monopoly pricing. Judge Posner has suggested another metric that is a refinement of the conventional deadweight loss analysis. In this Article, we review the current deadweight loss analysis of the social cost of monopoly. Most prominently we suggest three reasons to reconsider this analysis. First, the deadweight loss analysis uses the sum of consumer and producer surplus to give an approximate measure of gains and losses without giving any consideration to the consumers' relative utility levels. Second, the analysis relies on the questionable assumption of profit-maximizing firms, not taking into consideration that where the shares of firms are widely-held—as is the case with most firms that have monopoly—managers may be motivated by goals other than profit maximization. Third, the analysis is problematic to the extent that it ignores the social cost of inducing perfect competition—or alternatively, of increasing the output level to the socially optimal level—in a given industry, and thus assumes a counterfactual that is not attainable even by a benevolent social planner.

As an alternative approach to analyzing the social cost of monopoly, we propose an applied general equilibrium model. The index of social cost we use is the coefficient of resource utilization introduced by Gerard Debreu [Deb51]. This measure provides an exact, ordinal measure of the economic cost of monopolization in terms of wasted real resources. We take as benchmark a Pareto optimal state of economy that provides the same level of consumer satisfaction as achieved in the monopolized state.[7] The primary objective of antitrust policy is to promote consumer welfare and efficiency, and Pareto optimality embodies both of these objectives. To this extent, we suggest that *marginal cost pricing should be viewed not only as a consequence of perfect competition but also as a necessary condition for achieving Pareto optimality.*[8]

---

[7] See U.S.C. §18, Section III.A. Ultimately, our methodology can be used to calculate the snap-shot social cost in terms of the dollar values in relation to an attainable counterfactual.

[8] See U.S.C. §18, Part III. Whenever this condition is violated in a sequence of observed equilibria we can calculate the social cost in each observation via Debreu's coefficient of resource allocation. See U.S.C. §18, Section III.B.

The rest of this Article is divided into several sections. Section 2 is a review of the current analysis of the social cost of monopoly based on the deadweight loss triangle. In Section 3, we discuss how the notion of Pareto optimality as the benchmark state of the economy can lead to a more appropriate measure of the cost. In Section 4, we formalize the social cost analysis using a two-sector model and illustrate how to compute the coefficient of resource utilization. In Section 5, we derive a family of linear inequalities, where the unknown variables are utility levels and marginal utilities of income of households and the marginal costs of firms. These inequalities suffice for an empirical determination of monopoly power in a series of historical observations of market data, and this methodology can be used by courts in applying the *Grinnell* test.

2 Reconsidering the Deadweight Loss as the Social Cost of Monopoly

By now, most economists agree as to the *nature* of the problem posed by monopoly and market power. A monopolist who cannot price-discriminate has an incentive to reduce output and charge a price higher than marginal cost, and in turn, prevent transactions that would have been mutually beneficial. Faced with monopoly pricing, consumers either pay higher than necessary prices to obtain their goods or must choose false alternatives—alternatives that appear to be cheaper even though they might require more resources to produce.[9] Put differently, monopoly is inefficient because in preventing such transactions, society uses up more resources than necessary to achieve given levels of utility among consumers.[10] Although destruction of mutually beneficial transactions is patently inefficient from society's perspective, it remains unclear what is the proper metric to measure the social cost of monopoly. Intuition tells us that, whatever the metric is, it should indicate the extent to which the current state of monopolized economy deviates from an efficient state of economy that could have been achieved if resources were better allo-

---

[9] This point has long been recognized. Lerner in a landmark article [Ler37] writes that "[I]ncreasing the price of the monopolized commodity [causes] buyers to divert their expenditure to other, less satisfactory, purchases. This constitutes a loss to the consumer which is not balanced by any gain reaped by the monopolist, so that there is a net social loss."

[10] The foregoing analysis is the "resource allocation" aspect of monopoly. See Harberger [Har54]. Clearly, there is also a distribution effect: monopoly pricing tends to redistribute income in favor of the monopolist. But insofar as these are mere transfers, antitrust economists do not regard them as socially inefficient. See, e.g., Lerner [Ler37, p. 157] ("A levy which involves a mere transference to buyer to monopolist cannot be said to be harmful from a social point of view unless it can be shown that the monopolist is less deserving of the levy than the people who have to pay it...."); Posner [Pos01, pp. 13, 24].

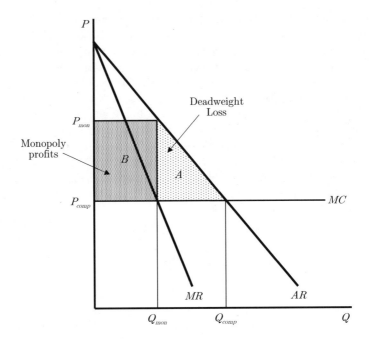

Fig. 1. Partial equilibrium analysis

cated. In traditional textbook microeconomics, the social cost of monopoly is measured by the deadweight loss triangle.

Triangle A in Figure 1 depicts this loss since this area represents the amount of additional social surplus that could have been realized had the pricing been at marginal cost. Alternatively, taking potential rent-seeking behavior among firms into consideration, Judge Posner [Pos75] argues that in certain markets where firms compete to become a monopoly, the social cost should include producer surplus in addition to the deadweight loss.[11] In Figure 1, this quantity is represented as the sum of A and B. Despite capturing the essence of the inefficiency of monopoly, the social cost metrics such as these, which are based on the conventional deadweight loss triangle, are inappropriate because they require some implausible assumptions.

First, these measures of social cost and surplus use the money metric: all benefits and inefficiencies are quantified in terms of dollars. How should we understand the relationship between the money metric and social surplus? Suppose Abigail is willing to pay as much as \$5 for a widget, and Brian, \$4,

---

[11] To formalize this idea, Posner sets forth three conditions that need to be satisfied for this assertion to hold: (1) firms compete to obtain a monopoly; (2) firms face a perfectly elastic long-run supply of all input; and (3) the costs the firms spend in attempting to obtain the monopoly serve no socially useful purposes. See also Posner [Pos01, pp. 13–17].

but a widget only costs $2. Then after purchasing a widget, Abigail is left with $3 to spare, and Brian, $2. They can devote their remaining dollars towards consumption of other goods. But to add these values together and say $5 is the measure of social welfare does not really tell us what benefit each of them could have derived from additional consumption; little information is revealed about consumer welfare. In order for this surplus measure to truly represent the social loss, we would need "the heroic assumption that a dollar is worth the same to everybody."[12] This notion of maximizing social surplus is related to the notion of Kaldor–Hicks efficiency but only loosely so since we are using dollar values as a proxy to measure social welfare. There may be instances where the only feasible and testable solution is to quantify all benefits and costs in terms of dollars. The cost-benefit analysis commonly used in health and safety regulation is one example. But the cost-benefit analysis paradigm, too, has been criticized on many grounds, not the least of which is the validity of this "heroic assumption."

A second failing of the deadweight loss analysis is that it relies on the concept of a profit-maximizing monopolist who produces goods until marginal cost equals marginal revenue. Let us forget for the moment that this directive may be extremely hard to carry out in reality due to imperfect information. What is somewhat striking is that even with perfect information, the profit-maximizing condition often fails to describe accurately actual behaviors of monopolists. The literature provides several reasons why a monopolist might not consciously seek to maximize profit. First of all, rarely do we see any prominent monopolist firm that has a sole owner or a sole shareholder. Instead, most monopolist firms have multiple shareholders; and in many cases, these firms' shares are widely held. Investment and cost decisions of a firm are ultimately made by the managers of the firm who receive salaries and bonuses but are not necessarily owners. These managers may neglect to maximize profits if that is not in their own best interest to do so. This could be the case, for instance, if managers were allowed to reap personal profits at the expense of corporate well-being; insider-trading is one example. The literature of industrial organization and microeconomic theory is replete with sources of inefficiency in principal-agent models. Fortunately, corporate law provides several institutional safeguards to minimize this type of opportunities for managers. Insider trading and false financial reporting are illegal under the Rules 16, 10(b), and 10(b)-5 of the Securities and Exchange Act of 1934. Managers and directors owe a duty of loyalty to their shareholders and cannot benefit themselves at expense of corporation. The corporate opportunity doctrine precludes directors from taking a business opportunity for their own when the opportunity is within the firm's line of business and the firm can afford to exploit the opportunity. And importantly, a majority of shareholders

---

[12] Posner [Pos01, p. 23]. In the jargon of economic theory consumer surplus only measures changes in consumers' welfare if the marginal utility of income is the same for every household, rich and poor alike.

can vote out the incumbent management in case they are dissatisfied with the firm's performance.

But more importantly, even if shareholders and corporate law can create incentive schemes for managers to induce them to do what is best for the shareholders, managers would maximize profits only if that is in the best interest of shareholders. Shareholders come in various types, and for many, the firm's profit-maximization can run afoul of their best interest.[13] Specifically, if a shareholder happens also to be a consumer of the firm's output, then she will suffer by paying high prices for the firm's output. Second, a shareholder who sells factor inputs to the firm would stand to lose if the firm uses its market power to drive down the price for the factor she sells. Third, a shareholder who owns a diversified portfolio may be hurt if the firm uses its market power to hurt its competitors. Finally, even if the shareholder has no interaction with any of the firm's output, she may be hurt if she consumes a good that is complementary to the firm's output, since the firm's pricing policy will necessarily impact the demand curves for complementary goods.

Non-economic arguments may also play a role, as "[w]hen the monopolist is not working on purely business principles, but for social, philanthropic or conventional reasons" or more likely "when the monopolist is working on purely business principles, but keeps the price and his profits lower than they might be so as to avoid political opposition" (Lerner [Ler37, p. 170]). Although the notion that a monopolist maximizes profit has some intuitive appeal, it may nevertheless run counter to the shareholder's best interests, and thus will not always be pursued. In actuality, a monopolist's behavior is more likely to resemble that of cost-minimization, rather than of profit-maximization. If monopolists do not obey the profit-maximizing conditions in practice, then the social cost of monopoly certainly should not be measured on the assumption that they do.

Our third and most important point is that the social cost of monopoly as measured by deadweight loss is problematic to the extent that it implicitly assumes that the relevant benchmark of efficiency—the counterfactual against which we measure the social loss—is the state of perfect competition. The rationale is that under perfect competition price will equal marginal cost, and a willing buyer and a willing seller will engage in transactions without wasting any resources. Nevertheless, the assumption of perfect competition as the ultimate benchmark is less innocuous than it appears. Perfect competition requires atomism of firms and buyers. But the literature is often silent as to exactly where these "other" firms suddenly come from. It is unlikely that there are these firms idly sitting around and not producing any socially usefully goods but waiting to enter this market. A more likely scenario is that

---

[13] This argument is only cursorily included in this section. For a more detailed treatment, see Kreps [Kre90]. As a result, Kreps concludes that the notion of profit-maximizing firm is more applicable to price-taking firms without market power.

firms or individuals somewhere have to cease their existing, socially useful activities in order to enter a particular industry. For this reason, the perfect competition benchmark unrealistically assumes a sudden costless creation of countless new firms while everything else in society remains unchanged. More generally, as A.P. Lerner [Ler37, p. 161] articulates, "[t]he direct comparison of monopolistic with competitive equilibrium ... assumes that cost conditions are the same and that demand conditions are the same. Neither of these is likely, and the combination of both is much less likely" (p. 161).[14]

We can illustrate the last point with a stylized example. Suppose we have a competitive market for widgets. But one day one of the widget-manufacturing firms, Firm A, patents a new formula to make "twidgets"—a new beneficial invention otherwise unrelated to widgets but a more profitable venture—out of the resources and technologies originally used to produce widgets. Twidgets become an instant hit in the market, but for the first twenty years, Firm A enjoys monopoly due to the patent right. We have a government-sanctioned inefficiency in this case. As the patent expires, other widgets companies will rush into the twidget market, consumers can then buy twidgets at marginal cost and the deadweight cost from the widget monopoly is extinguished. Notice, however, that the widget market will likely suffer now because widget firms are expending their resources to producing twidgets. In other words, in order to induce perfect competition in one industry that was originally monopolistic, one would have to pull out resources from other industries. The supply curve will, then, shift inward in one or more of other industries, and the new resource allocation reduces the social surplus generated from those industries.

Figures 2 and 3 demonstrate this example. Figure 2 represents the twidget market, and Triangle ABC measures the gain in social surplus due to competition in the twidget market. Meanwhile, Figure 3 refers to the competitive widget market and the counterfactual when the twidget patent expires. Area represented by $EFGH$ measures the reduction in social surplus due to the inward shift of the supply curve in the widget market. A more accurate measure of the social cost, therefore, would have to consider the totality of circumstances; the gain ABC would have to be measured against the loss $EFGH$. All of sudden, it is not at all obvious that eliminating deadweight loss by inducing perfect competition in the twidget industry is particularly desirable; the result may be overall reduction in social surplus.

An economic analysis which focuses on the social surplus of one sector without considering possible implications for other sectors is called partial equilibrium analysis. Partial equilibrium analysis remains a powerful methodology for analyzing the behavior of firms in an isolated market where the impact on prices in other markets is negligible. And yet this is hardly the

---

[14] Lerner is making one additional observation that the long-run cost curve faced by a single firm in general may not be the same long-run cost curve if many firms are competing in the market.

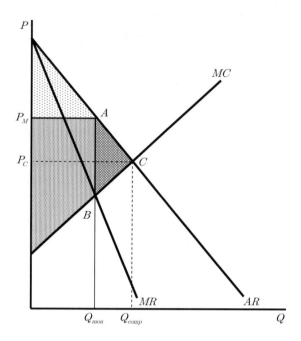

Fig. 2. Monopoly and competition in the twidget market

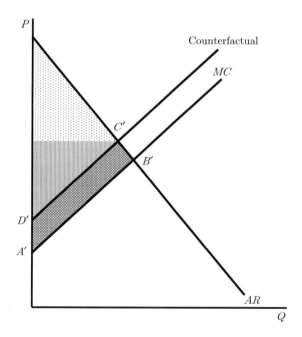

Fig. 3. Competitive widget market and counterfactuals

case with interesting instances of monopoly power, e.g., AT&T, IBM, and Microsoft. In all of these cases, prices were affected well beyond the immediate markets, and the static one-sector model cannot correctly estimate the social cost of monopoly. At a minimum, we must consider the effect of inducing perfect competition in one industry on a different industry from which resources are drawn; a proper model thus would have to consider at least two separate sectors with common factors (which can be, broadly speaking, capital and labor). What is more, since inducing perfect competition in and of itself may not be desirable on the whole, we must also consider other states of the economy that are potentially superior. Finally, even if perfect competition only serves as a proxy for marginal cost pricing, the marginal cost pricing counterfactual, too, is no more appropriate as a benchmark of welfare comparison. The marginal cost pricing necessarily requires a higher output level than the given state of monopoly, and in turn, requires more resources, including capital and labor, to be drawn from other industries.

3 Consumer Welfare, Pareto Optimality and General Equilibrium Theory

In general equilibrium theory, consumers simultaneously provide labor and capital to firms, own shares of the firms, and maximize their utility based on consumption subject to their income constraints; meanwhile, firms from different sectors produce different goods but use common factor inputs, labor and capital. Notwithstanding the seemingly all-encompassing features of general equilibrium theory, its application to antitrust policy and the social cost of monopoly has been remarkably limited to date.[15] In this section, we briefly discuss the elements of general equilibrium theory and propose an alternative method of measuring the social cost of monopoly power—one not subject to the concerns raised above but nonetheless consistent with the aims of antitrust law.

We stress that a reasonable measure of the social cost of monopoly should be based on a proper counterfactual. If perfect competition is inappropriate as a counterfactual, then what should be the ideal state of the economy against which to measure the social cost of monopoly? We propose that the benchmark of comparison for the purpose of measuring the social cost of monopoly should be a counterfactual *state that achieves the same or greater level of utility for everyone but with the least amount of resources.* If such a state can be constructed, then the economic cost of monopoly is simply the dollar amount of wasted resources; the given monopolized state performs no better than the counterfactual for any individual but simply uses up more resources.

---

[15] The conspicuous absence of application of general equilibrium theory to antitrust law is due in part to the indeterminacy of the price level in the Arrow–Debreu general equilibrium model. As such, the model does not admit price-setting, profit-maximizing firms.

The proposal merits some explanation. The notion of efficiency and welfare in general equilibrium theory is Pareto optimality, also known as *allocative efficiency*.[16] A state of the economy is said to be Pareto optimal if no consumer can be made better off by reallocating productive resources and engaging in mutually beneficial trades without making another consumer worse off; Pareto optimality thus represents a state of maximal consumer welfare. The crown jewels of general equilibrium theory are the two welfare theorems. The first welfare theorem states that every competitive equilibrium—i.e., equilibrium achieved under perfect competition—is Pareto optimal (for an excellent treatment of general equilibrium, see Mas-Colell et al. [MWG95]). In short, we tend to "value competition because it promotes efficiency, that is, as a means rather than as an end" (Posner [Pos01, p. 28].[17]

The primary goals of antitrust policy are efficiency and the enhancement of consumer welfare.[18] Both of these concepts appeal to Pareto optimality. Robert Bork "[insisted] that the achievement of Pareto optimality was the sole objective of Congress (as long as 1890) when it enacted the nation's antitrust statutes" [Hyl03, p. 5]. Similarly, President Reagan's first Council of Economic Advisers specifically defined efficiency in an economy in terms of Pareto optimality, not Kaldor–Hicks efficiency.[19] Curiously, this nexus between Pareto optimality and antitrust law has been all but overlooked in the economics literature due to the singular focus on the deadweight loss analysis.[20] Our analysis restores this nexus and suggests that the proper benchmark for measuring the cost of monopoly should be a Pareto optimal state of the

---

[16] The literature appears to use these terms interchangeably. See, e.g., Morgan [Mor00] (describing allocative efficiency as the state in which "there [is] no combination of production or exchange that could make anyone better off without making someone else worse off").

[17] In addition, Lerner [Ler37, p. 162] notes that the "importance of the competitive position lies in its implications of being ... the position in which the 'Invisible Hand' has exerted its beneficial influences to the utmost."

[18] William F. Baxter, the first antitrust chief during the Reagan Administration, says "the antitrust laws are a 'consumer welfare prescription'—that is, they are intended to promote economic efficiency, broadly defined." Likewise, Bork insists that the "only legitimate goal of American antitrust law is the maximization of consumer welfare...." See Adams et al. [AB91]. Modern courts also appear to understand the aim of antitrust law as enhancement of consumer welfare. Case law suggests that courts by way of antitrust law have shifted their focus from promoting competition to maximizing consumer welfare. See Hylton [Hyl03, p. 40].

[19] According to the Council, an economy "is said to be 'efficient' if it is impossible to make anyone better off without making someone else worse off. That is, there is no possible rearrangement of resources, in either production or consumption, which could improve anyone's position without simultaneously harming some other person." See Hylton [Hyl03].

[20] However, one of the few analyses that link monopoly and Pareto optimality was given by Lerner [Ler37] as early as 1937.

economy, not simply competitive markets. A correct social cost metric should therefore reflect both the degree of deviation from a Pareto optimal state of the economy and the dollar amounts wasted.

But an economy can have possibly infinitely many Pareto optimal allocations, and we have not yet specified which of the many possible Pareto optimal states we should take as the relevant benchmark. We propose the unique Pareto optimal state characterized by Debreu's *coefficient of resource utilization*, ρ. This coefficient is the smallest fraction of total resources capable of providing consumers with utility levels at least as great as those attained in the monopolized economic state.[21] Hence the efficiency loss in real terms is $(1 - \rho) \times$ total resources; the economy can throw away $(1 - \rho) \times$ total resources and not make anyone worse off.[22] For example, suppose Abigail has ten apples, Brian has ten pears, and although each would prefer a mixed bundle of pears and apples, they are prohibited from trading for some reason. If Abigail is indifferent between having ten apples and having a bundle of two apples and four pairs, and Brian between ten pears and a bundle with five apples and three pears, then the current state of economy is no better off than one that could be achieved with only seven apples and seven pears. Thus society is squandering three apples and three pears, as they add nothing to consumer welfare. If no smaller bundle can achieve the same level of consumer welfare as the current state of economy, then the coefficient of resource utilization in this case is 0.7.

Importantly, the fact that we choose ρ to be the minimal coefficient renders the new state of the economy—in which nobody is worse off than in the monopolized state Pareto optimal relative to the reduced resource endowment. Recall our discussion earlier that the inefficiency of monopoly could be viewed as using up more than necessary amounts of resources to achieve a particular utility level. Then the natural benchmark for a monopolized economy is the Pareto optimal economic state that uses the least amount of resources but produces the same or higher level of consumer satisfaction; specifically, society's endowment in the new state will be exactly $\rho \times$ total resources. The

---

[21] Debreu [Deb51] analyzes economic loss associated with nonoptimal economic states and identifies three kinds of inefficiencies in economic systems. Only one need concern us here: "imperfection of economic organization such as monopolies or indirect taxation or a system of tariffs." To measure the economic loss, he posits a cost-minimization problem dual to Pareto's maximization of social welfare.

[22] An attentive reader might reason that this measure actually offers a lower bound of the social cost since applying the same ρ across all resources is constraining. Indeed, if we can determine the minimum level of resources necessary to achieve the same utility level without imposing the same proportion of reduction across all resources, the measure of social cost might be greater. Nevertheless, there is no guarantee that such minimum bundle of resources is uniquely determined. The coefficient of resource utilization provides the benefit that such a level necessarily exists and is uniquely determined.

associated economic cost indicates the inefficiency due to monopolization and can be converted into a dollar amount.

This cost is indicated in Figure 4.[23] The original production possibility frontier ("PPF") can be thought of as a social budget constraint. α_1 is the given state of economy, and α_2 is an alternative state that lies on the community indifference curve and is tangent to the counterfactual PPF. The counterfactual PPF represents the PPF produced with "minimal" social resources and yet is tangent to the community indifference curve, meaning every individual in the community is indifferent between the current state and the counterfactual state.

Of course, this notion of economic cost would have meaning only insofar as the relevant benchmark is actually achievable. After all, one of the reasons for which we are not satisfied with the deadweight loss triangle as the measure of the cost of monopoly was that the counterfactual was not an achievable state of economy. This is where the second welfare theorem of general equilibrium theory comes in, which tells us that *every* Pareto optimal economic state can be realized as a competitive equilibrium with lump sum transfers of income between households (see Mas-Colell et al. [MWG95]). As a result, using only $\rho \times$ total resources and with lump sum transfers, society can achieve the desired Pareto optimal state. Hence a benevolent social planner with perfect

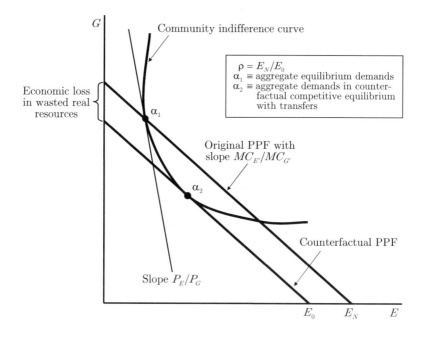

Fig. 4. General equilibrium analysis

---

[23] We thank T.N. Srinivasan for suggesting this diagram.

information could achieve our counterfactual while he cannot achieve the perfect competition benchmark. In addition, the resulting measure of social cost provides the added benefit of assuming only ordinal measures of utility. We need no longer assume either that a dollar is worth the same to everyone or that utility functions can be aggregated across consumers. Adding up the cost of resources too makes sense with this framework since we are not equating these economic costs with gains or losses in individual utility levels. Our task thus reduces to estimating ρ and the amount of resources wasted in a given monopoly state.

Before we go on, we want to make an important observation. Our exclusive focus on Pareto optimality seems to ignore the welfare of the producers, and this may appear to be inconsistent with the original concern for social surplus, which includes both consumer surplus and producer surplus. Not so. If one were to make an analogy at all, general equilibrium theory's notion of Pareto optimality should be compared with partial equilibrium theory's notion of social surplus, not consumer surplus. This is because in general equilibrium theory, consumers own the firms, and thus the firm's profits are distributed back to the consumers according to their shares of ownership. Firm decisions are made so as to maximize the welfare of consumer–shareholders, and there are no personas associated with the producers.

4 A Two-Sector Model and Cost-minimizing Equilibria

We now formalize our idea and illustrate the computation of ρ with a two-sector general equilibrium model.[24] By now, we hope the motivation for having a two-sector model is clear to the readers: if the cost of monopoly in an industry can only be analyzed in relation to another industry that shares the same factor inputs, a proper analysis of the social cost must include at least two sectors. While our model can be generalized to accommodate multiple sectors, we only really need two sectors in order to convey the main ideas effectively. There are two consumers, two commodities, two firms, and two factors of production.[25] We make the standard assumptions from microeconomic theory. Consumers have smooth, concave monotone utility functions

---

[24] We formulate our model following Shoven and Whalley's work [SW92]. The relevant parts are Sections 3.2 and 3.3, and also Chapter 6, where the authors include Arnold Harberger's two-sector general equilibrium analysis of capital taxation. This model can easily extend to three or more sectors.

[25] This model is widely used in the applied fields of international trade and public finance where the focus is on general equilibrium comparative statics for policy evaluation. Also the data available such as national accounts and input-output data are easily accommodated in a two-sector model. For readers unfamiliar with the properties of the two-sector model, we recommend Shoven and Whalley [SW92]. Here we follow the notation in Brown and Heal [BH83].

and endowments of capital and labor and shareholdings in firms; they maximize utility subject to their budget constraints and are price-takers in the
product and factor markets. Each firm produces a single output with a smooth
monotone and strictly quasi-concave production function. In equilibrium all
markets clear.

Competitive firms maximize profits and are price-takers in both the product and competitive factor markets; they produce at minimum cost and price
output at marginal cost. Most of general equilibrium theory conventionally
assumes competitive markets in all sectors.[26] In order to extend the paradigm
to encompass the existence of monopoly power, we introduce a new notion
of market equilibrium: firms with monopoly power have unspecified price-
setting rules for output—where the price of output is a function of the level
of the output (see Kreps [Kre90])—but are assumed to be cost-minimizing
price-takers in competitive factor markets.[27] This means, for instance, that
Microsoft may have monopoly power in the software market but it still needs
to pay competitive wages for its employees. Meanwhile, in equilibrium they
make supra-competitive profits since the monopoly price exceeds the marginal
cost of production. Our analysis derives from a subtle but important distinction between price-setting profit-maximization—which we rejected—and
monopoly power, i.e., the power to raise price above the competitive level and
make supra-competitive profits.[28] Both OPEC and Microsoft have monopoly
power under this definition and it seems reasonable to assume that both attempt to produce output at minimum cost. But neither OPEC nor Microsoft
appears to be setting prices to maximize monopoly profits.

We denote the two consumers as x and y. The inputs or factors are capital
(K) and labor (L). The outputs or goods are natural gas (G) and electricity
(E). Each consumer has a utility function denoted U_x and U_y. Consumers are
endowed with capital and labor, which they provide to firms in exchange for
wages and rents; they also have shares in the ownership of the firm. Endowments and shareholdings in firms for x and y are given by (K_x, L_x), (K_y, L_y);
$(\theta_{xG}, \theta_{xE})$, $(\theta_{yG}, \theta_{yE})$. Each firm has a production function, F_G and F_E. Let
$K = K_x + K_y$, and $L = L_x + L_y$. Let P_G and P_E denote the prices of natural
gas and electricity, and w and r denote the prices of labor and capital. Consumers can freely trade goods with each other, but not their labor or capital
endowment; firms can freely trade factor inputs. We suppose that the gas
market is competitive but the electricity market is monopolized. Therefore

---

[26] In fact, the Arrow–Debreu general equilibrium model, with its indeterminate
absolute price level, cannot accommodate price-setting, profit-maximizing firms.
See Cornwall [Cor77] and Bohm [Boh94].

[27] "By pure monopoly is meant a case where one is confronted with a falling demand
curve for the commodity one sells, but with a horizontal supply curve for the
factors one has to buy for the production of the commodity; so that one sells as
a monopolist but buys in a perfect market" (Lerner [Ler37])

[28] This latter definition is consistent with Lerner's index [Ler37]. "If P = price and
C = marginal cost, then the index of the degree of monopoly power is (P – C)/P."

$P_G = MC_G$, the marginal cost of producing gas, and gas is produced with constant returns to scale.

Let us consider how the economy operates. Consumers consume electricity and gas to maximize their utility subject to their budget constraints. They have several sources of income: wages from providing labor, interest on their capital investment, and dividends from the firms' shares, which are determined by the firms' profits. Therefore, we write the consumer's problem as follows:

Consumer's Problem

$$\max U_i(E_i, G_i) \text{ subject to } P_E E_i + P_G G_i \leq I_i \tag{1}$$

where $I_i = wL_i + rK_i + \theta_{iG}(P_G G - wL_G - rK_G) + \theta_{iE}(P_E E - wL_E - rK_E)$ and $i = x, y$.

Since utility increases in E_i and G_i, the weak inequality ends up binding. In addition, since the gas market is competitive, the third term in the income equation is zero. Meanwhile, firms minimize their cost of production given their target levels of production.

Firm's Problem

$$\min wL_j + rK_j \text{ subject to } F_j(L_j, K_j) = j, \text{ for } j = E, G. \tag{2}$$

Notice that these target levels are not necessarily determined by profit-maximization motives. For monopoly or any other market structure it matters not how the actual target levels are chosen; our methodology gives a measure of inefficiency based on the observable production levels the firms choose and market prices.

Now we give our definition of cost-minimizing market equilibrium.

Definition. A *cost-minimizing equilibrium* is then defined as a set of relative prices P_E/w, P_G/w and r/w; consumer's demands for goods E_x, G_x and E_y, G_y; firm's demands for factors L_E, K_E and L_G, K_G; and output levels E and G such that (i) consumers maximize their utility levels given the prices of goods, (ii) firms make nonnegative profits and minimize their costs of production given the prices of factor inputs; and (iii) all markets clear. That is,

(1) Product Markets: $E_X + E_Y = E$; $G_X + G_Y = G$
(2) Factor Markets: $L_E + L_G = L$; $K_E + K_G = K$
(3) Nonnegative Profits: $P_E E \geq wL_E + rK_E$; $P_G G = wL_G + rK_G$.

An important result from general equilibrium theory is the set of conditions necessary for Pareto optimality of the economy.[29] We will first state them and explain intuitively why these conditions are necessary:

---

[29] For the intuition and derivation of these necessary conditions, see generally Bator [Bat57].

$$MRS_x = MRS_y \tag{3}$$

$$MRTS_E = MRTS_G \tag{4}$$

$$MRS_x = MRT = MC_G/MC_E. \tag{5}$$

To begin with, what can we say about a Pareto optimal state of the economy? At an optimal state, consumption should be efficient in the sense that the consumers should not be able to trade their consumption goods with each other and achieve a Pareto improvement. In addition, production should be efficient in the sense that the firms should not be able to trade their factor inputs and achieve a Pareto improvement on society's production levels. And finally, we also want to make sure the product-mix is efficient in the sense that society should not elect to produce a unit of gas instead of some additional amount of electricity if a consumer prefers more electricity to gas in his consumption. The above conditions are simply abstractions of these intuitions.

MRS_x refers to the marginal rate of substitution of electricity for gas for x, and it represents the rate at which x is willing to give up electricity for gas, holding his utility constant. This is also equal to the ratio of marginal utilities for each good: $MU_{x,G}/MU_{x,E}$. The first condition says that at optimum, x's willing rate of substitution must equal that of y's. Let us see why this is true. Without loss of generality, suppose that $MRS_x = 2$ and $MRS_y = 1$ at some point. This state cannot be optimal. x is willing to give up as much as 2 units of electricity to obtain 1 unit of gas, and y is willing to make a one-to-one trade and still able to maintain her current utility level. Then y can choose to trade one unit of her gas to extract two units of electricity from x. This exchange will not change x's utility but will increase y's utility since y would have achieved the same level of utility with just one unit of electricity and now she ends up with one extra unit. Since this is a strict improvement for y without hurting x, the new state is a Pareto improvement to the original state, contradicting our assumption that the current state is Pareto optimal. Thus we must have $MRS_x = MRS_y$.

The second condition refers to the marginal rate of transformation, and the analysis is similar to the first one. The marginal rate of technical substitution measures the rate at which the firm can replace one input, say labor, by the other, capital while maintaining the same production level. If $MRTS_E = 2$, this means, Firm E can produce the same amount of electricity while trading in two units of labor for one unit of capital. It is then easy to see why we need $MRTS_E = MRTS_G$ at optimum. For example, if $MRTS_E = 2$ and $MRTS_G = 1$, then Firm E can maintain its current production level by taking one additional unit of capital and giving up two units of labor. Since Firm G can trade at a one-to-one ratio and maintain its current level of production, Firm G can increase its production level by offering one unit of its capita to Firm E and receiving two units of labor. This is an overall improvement to the current state, and thus it violates the optimal condition. Therefore, we must have $MRTS_E = MRTS_G$ at optimum.

The third condition equates MRS_x with MRT, the marginal rate of transformation. MRT represents how many units of electricity must be sacrificed in order for society produce gas; this incorporates the marginal costs of production for both. That MRT should equal to the ratio of MC_G and MC_E can be explained by the fact that is MC_G the cost to society of producing one additional unit of gas (by expending some combination of labor and capital) and MC_G the cost to society of producing one additional unit of electricity. A more interesting question is why MRS_x should equal MRT. If $MRS_x = 2$ but $MRT = 1$, for example, that means x is willing to give up as much as 2 units of electricity to obtain 1 unit of gas. Since the costs to society are equal for production of gas and production of electricity at the margin, it would have been better to have forgone the production of the last unit of electricity and instead devote this resource to producing an additional unit of gas. This would have made x happier since his utility level would have been the same with giving up two units of electricity and obtaining one unit of gas, but with society's alternate production plan he need only give up one unit of electricity and obtain one unit of gas. Therefore, we need $MRS_x = MRT$. Analogously, $MRS_y = MRT$ and in the end society's marginal rate of transformation must be equal to the marginal rate of substitution for every consumer in the economy.

In addition to these necessary conditions, we derive a few more conditions from the consumers' and the firms' optimization problems. The firm's cost minimization problem relates the marginal rate of technology substitution with wages and rental rates. As for the consumer's problem, since the consumers make their consumption decisions based on the market prices, the first-order condition from the consumers' problem tells us that:

$$MU_{x,G}/MU_{x,E} = P_G/P_E. \tag{6}$$

Since $MRS_x = MU_{x,G}/MU_{x,E}$, if we combine (6) with (5), we have

$$P_G/P_E = MC_G/MC_E. \tag{7}$$

And $P_G = MC_G$, since the market for gas is competitive. Hence for Pareto optimality, we must also have

$$P_E = MC_E. \tag{8}$$

It is in this sense that we should view marginal cost pricing not only as a result of perfect competition but also as a necessary condition for society to achieve Pareto optimality.

We now turn to the computation of ρ in this two-sector model. Suppose the given economic state of the model is a cost minimizing market equilibrium where $P_E/P_G \neq MC_E/MC_G$, and suppose in equilibrium x consumes $(\hat{E}_x, \hat{G}_x)$ and y consumes $(\hat{E}_y, \hat{G}_y)$. ρ is the minimum α between 0 and 1 where the given two-sector model with reduced social endowments αK and αL can produce sufficient electricity $\check{E}$ and natural gas $\check{G}$ such that:

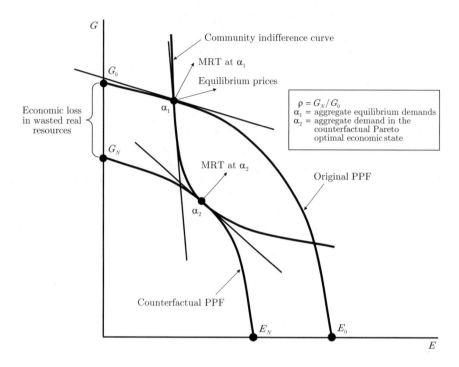

Fig. 5. The social cost of monopoly

$$U_x(\breve{E}_x, \breve{G}_x) \geq U_x(\hat{E}_x, \hat{G}_x) \tag{9}$$

$$U_y(\breve{E}_y, \breve{G}_y) \geq U_y(\hat{E}_y, \hat{G}_y) \tag{10}$$

$$\breve{E}_x + \breve{E}_y = \breve{E}; \quad \breve{G}_x + \breve{G}_y = \breve{G} \tag{11}$$

$$\breve{E} = F_E(\underline{L}_E, \underline{K}_E); \quad \breve{G} = F_G(\underline{L}_G, \underline{K}_G) \tag{12}$$

$$\underline{L}_E + \underline{L}_G = \alpha L; \quad \underline{K}_E + \underline{K}_E = \alpha K. \tag{13}$$

These equations and inequalities define the optimization problem for determining ρ and we can solve them using the Lagrange multiplier method.

We can illustrate this with Figure 5. The outputs $(\hat{E}, \hat{G})$ produced in a cost minimizing market equilibrium lie on the PPF, as a consequence of competitive factor markets and production at minimum cost. α_1 is the output $(\hat{E}, \hat{G})$ produced in the cost minimizing market equilibrium. $\alpha_2 = (\breve{E}, \breve{G})$ and satisfies (9)–(13). The social endowments used to produce α_2 are ρK and ρL, where K and L are the original social endowments of capital and labor. If the slope of the PPF is P_E^*/P_G^*, then ρ is the ratio G_N/G_0 and the economic cost is $(1-\rho)[wL+rK]$.[30] The existence of such α is guaranteed since solutions to

---

[30] This computation is given in Brown and Wood [BW04].

(9)–(13) are guaranteed for $\alpha = 1$ by virtue of the existing market allocations. The uniqueness is also assured since we choose the minimal such α. In practice, ρ must be estimated from market data, and in Section 6 we show how this is done.

A natural question at this point is how this dollar amount would compare to the dollar amount of the deadweight loss triangle. As it turns out, there are no systematic relationships between these two figures. The economic wastes from the applied general equilibrium model can be lower, higher, or equal to the dollars corresponding to the deadweight loss costs.

Before we conclude this section, we make a side remark. We showed above thatmarginal cost pricing was a necessary condition for Pareto optimality. Supra-competitive pricing, of course, is not the only instance where (8) is violated: firms practicing predatory pricing violate (8) also by artificially setting prices below marginal costs. Thus, this perspective on marginal cost pricing illuminates an important aspect of predatory pricing: *the harm in predatory pricing is that by selling goods at a price below marginal cost, the firm destroys Pareto optimality in society in much the same way monopoly pricing does.* This observation challenges the current approach towards predatory pricing in antitrust law established in *Brooke Group Ltd. v. Brown & Williamson Tobacco Corp.*[31] Under this standard, an incumbent monopolist cannot be held liable for predatory pricing unless plaintiff can show not only that the monopolist priced goods below marginal cost but also that the monopolist had a reasonable prospect of recouping the incurred costs. Judge Easterbrook reasoned in another case that "if there can be no 'later' in which recoupment could occur, the consumer is an unambiguous beneficiary even if the current price is less than the [marginal] cost of production."[32] Our model shows that a monopolist who practices predatory pricing incurs social cost even absent the prospect of driving out competition or the prospect of recouping the costs. This symmetry between monopoly pricing and predatory pricing should not come as a surprise in light of the fact that predatory pricing, too, offers consumers false alternatives in terms of consumption goods, just as monopoly pricing does. Due to the lowered pricing, consumers may elect to consume a particular good over another even though the consumed good may be more costly to produce.

5 Application to the *Grinnell* Test

In this section, we consider how we can implement our model to measure monopoly power. One definition of monopoly power in the literature of antitrust economics is the existence of substantial market power for a significant

---

[31] 509 U.S. 209 (1993).

[32] A.A. Poultry Farms, Inc. v. Rose Acre Farms, Inc., 881 F.2d 1396, 1401 (7th Cir. 1989).

period of time. For cases involving Section 2 of the Sherman Act, courts use the *Grinnell* test: the offender must have both "(1) the possession of monopoly power in the relevant market and (2) the willful acquisition or maintenance of that power as distinguished from growth or development as a consequence of a superior product, business acumen, or historic accident."[33] The determination of the second element depends on the intent of the monopolist and will necessarily turn on the factual background of the case; the court will have to look at the business practice and exclusionary conduct. The first element, however, is an empirical question and its determination must turn on the history of price and demand data over a period of time. That is, in order for the courts to apply the *Grinnell* test they must review a history of the alleged monopolist's pricing behavior to ascertain the existence of monopoly power. The difficulty with inference of monopoly power is that neither the cost curves nor the demand curves are generally known; market data only provide us with the equilibrium behaviors of consumers and firms. Nonetheless we can combine some of the results from advanced microeconomic theory with our model to determine the existence of monopoly power.

Let us consider how this would work and what it is that we want to find. If a given industry is relatively competitive, then the price will be close to the marginal cost, the social cost from the firms' behavior will be small and the resulting ρ will be close to 1. If we can calculate ρ and find that it is significantly smaller than 1, then this is evidence that the market is not competitive, and we can infer monopoly power accordingly. But in order to calculate ρ, we must solve the minimization problem defined by Equations (9)–(13) and we have neither the utility functions nor the firm's production functions to work with. Instead we only have a history of market data, which tells us how consumers' behaviors[34] have changed over time with firms' varying prices and how firms' production levels have changed over time with varying factor prices. Since these data only provide us with the equilibrium behaviors, they are incomplete in that the utility functions of consumers and the production functions of firms are not observable. Microeconomic theory tells us how we can use observed equilibrium consumptions, factor demands of firms and market prices to approximate these functions from the equilibrium inequalities. At this point we can solve the problem defined by (9)–(13) and impute the economic costs of monopolization in terms of Debreu's coefficient of resource utilization.

The equilibrium inequalities consists of: the Afriat inequalities for each consumer; her budget constraint in each observation; Varian's cost minimizing inequalities for each firm; the market clearing conditions for the goods and factor markets in each observation; and the nonnegative profit conditions for each firm in each observation.[35] The Afriat inequalities follow from the Consumer's Problem in (1) and consist of a finite number of linear inequali-

---

[33] United States v. Grinnell Corp., 384 U.S. 563, 570-71 (1966).

[34] We assume these data are available at the household level.

[35] For discussions on Afriat's and Varian's inequalities, see Afriat [Afr67] and Varian [Var82]. The Afriat's inequalities and the Varian's inequalities have been em-

ties derived from a finite number of observations on a consumer's demands. Suppose we are given a history of demand pattern $\mathbf{x}_1, \mathbf{x}_2, ..., \mathbf{x}_n$ at market prices $\mathbf{p}_1, \mathbf{p}_2, ..., \mathbf{p}_n$. Each $\mathbf{x}_i$ is a bundle of goods at the household level, and each $\mathbf{p}_i$ is a vector of prices. We say that the Afriat inequalities are solvable if there exist a set of utility levels and marginal utilities of income for each observation, $\{(V_1, \lambda_1), ..., (V_n, \lambda_n)\}$, such that given any pair of observations, i and j, we have

$$V_i \leq V_j + \lambda_j \mathbf{p}_j \cdot (\mathbf{x}_i - \mathbf{x}_j) \quad \text{and} \quad V_j \leq V_i + \lambda_j \mathbf{p}_i \cdot (\mathbf{x}_j - \mathbf{x}_i). \quad (14)$$

Afriat's celebrated theorem is that if we can find the solutions to these V_i and λ_i, then we can find a concave, monotonic and continuous utility function $U(x)$ such that $U(\mathbf{x}_i) = V_i$ and rationalizes the data.

Varian's cost minimizing inequalities use a similar concept. They follow from the Firm's Problem from (2) and consist of a finite number of linear inequalities derived from a finite number of observations on a firm's outputs $f_1, f_2, ..., f_n$; factor demands: $\mathbf{y}_1, \mathbf{y}_2, ..., \mathbf{y}_n$; and factor prices: $\mathbf{q}_1, \mathbf{q}_2, ..., \mathbf{q}_n$. Varian showed that if we can find a set of numbers, $\{\beta_1, ..., \beta_n\},$[36] such that given any pair of observations, i and j, we have

$$f_i \leq f_j + \beta_j \mathbf{q}_j \cdot (\mathbf{y}_i - \mathbf{y}_j) \quad \text{and} \quad f_j \leq f_i + \beta_i \mathbf{q}_i \cdot (\mathbf{y}_j - \mathbf{y}_i), \quad (15)$$

then we can construct a continuous, monotonic, quasi-concave cost function such that the firm's decisions are consistent with the cost-minimization problem.[37]

We can combine these results with our model to estimate ρ from market data. Given a history of observations on the two-sector model, the equilibrium inequalities are solvable linear inequalities in the utility levels and marginal utilities of households and the marginal costs of firms, for parameter values given by the observed market data—that is, market prices, factor endowments, consumption levels, and share holdings in firms—if and only if this is a history of cost minimizing market equilibria. The solution determines a utility function for each household and a production function for each firm that is consistent with the market data in each observation. Using these utility functions and production functions we can solve the minimization problem for ρ defined by equations (9)–(13).

---

pirically tested in financial markets. See Bossaerts et al. [BPZ02] and Jha and Longjam [JL03].

[36] β_i turns out to be the reciprocal of the marginal cost at each observation.

[37] A production function f is said to rationalize the data if for all i, $f_i(\mathbf{y}_i) = f_i$ and $f_i(\mathbf{y}) \geq f_i(\mathbf{y}_i)$ implies $\mathbf{q}_i^i \cdot \mathbf{y} > \mathbf{q}_i^i \cdot \mathbf{y}_i$. That is, y minimizes the cost over all bundles of factors that can produce at least f_i. Varian's theorem is that the cost minimizing inequalities are solvable if and only if there exists a continuous monotonic quasi-concave, i.e., diminishing marginal rate of substitution along any isoquant, function that rationalizes the data. See Varian [Var82].

6 Conclusion

In this Article, we join with Robert Bork and William Baxter in proposing Pareto optimality as the embodiment of the goals of antitrust law. As such, it implicitly defines the proper benchmark for assessing the social cost of monopoly as the Pareto optimal state that utilizes minimal economic resources to provide the same level of consumer satisfaction as realized in the monopolized state. These wasted real resources provide a measure of the social cost of monopoly free from the vagaries of the social surplus measure used in conventional deadweight loss analysis of monopoly pricing, such as assuming a constant and equal marginal utility of income across consumers. Our model uses applied general equilibrium theory, which allows for the effects of monopolization on multiple sectors in the economy, an empirical determination, in a series of historical observations, of allegations of monopoly power, as required by the *Grinnell* test, and a reappraisal of predatory pricing.

Acknowledgments

The authors would like to thank G.A. Wood, Al Klevorick, Paul McAvoy, Ian Ayres, Keith N. Hylton, T.N. Srinivasan, and Eric Helland for their helpful comments on earlier drafts.

   Lee, Y.A., Brown, D.J.: Competition, consumer welfare and the social cost of monopoly. In: Collins, W.D. (ed) Issues in Competition Law and Policy. American Bar Association Books, Washington, DC (2007). Reprinted by permission of the American Bar Association.

Two Algorithms for Solving the Walrasian Equilibrium Inequalities

Donald J. Brown[1] and Ravi Kannan[2]

[1] Yale University, New Haven, CT 06520 donald.brown@yale.edu
[2] Yale University, New Haven, CT 06520 ravindran.kannan@yale.edu

Summary. We propose two algorithms for deciding if the Walrasian equilibrium inequalities are solvable. These algorithms may serve as nonparametric tests for multiple calibration of applied general equilibrium models or they can be used to compute counterfactual equilibria in applied general equilibrium models defined by the Walrasian equilibrium inequalities.

Key words: Applied general equilibrium analysis, Walrasian equilibrium inequalities, Calibration

1 Introduction

Numerical specifications of applied microeconomic general equilibrium models are inherently indeterminate. Simply put, there are more unknowns (parameters) than equations (general equilibrium restrictions). Calibration of parameterized numerical general equilibrium models resolves this indeterminacy using market data from a "benchmark year"; parameter values gleaned from the empirical literature on production functions and demand functions; and the general equilibrium restrictions. The calibrated model allows the simulation and evaluation of alternative policy prescriptions, such as changes in the tax structure, by using Scarf's algorithm or one of its variants to compute counterfactual equilibria. Not surprisingly, the legitimacy of calibration as a methodology for specifying numerical general equilibrium models is the subject of an ongoing debate within the profession, ably surveyed by Dawkins et al. [DSW00]. In their survey, they briefly discuss multiple calibration. That is, choosing parameter values for numerical general equilibrium models consistent with market data for two or more years. It is the implications of this notion that we explore in this paper.

Our approach to counterfactual analysis derives from Varian's unique insight that nonparametric analysis of demand or production data admits extrapolation, i.e., "given observed behavior in some economic environments, we can forecast behavior in other environments," Varian [Var82, Var84]. The forecast behavior in applied general equilibrium analysis is the set of counter-factual equilibria.

Here is an example inspired by the discussion of extrapolation in Varian [Var82], illustrating the nonparametric formulation of decidable coun-terfactual propositions in demand analysis. Suppose we observe a consumer choosing a finite number of consumption bundles x_i at market prices p_i, i.e., $(p_1, x_1), (p_2, x_2), ..., (p_n, x_n)$. If the demand data is consistent with utility max-imization subject to a budget constraint, i.e., satisfies GARP, the generalized axiom of revealed preference, then there exists a solution of the Afriat inequal-ities, U, that rationalizes the data, i.e., if $p_i \cdot x \le p_i \cdot x_i$ then $U(x_i) \ge U(x)$ for $i = 1, 2, ..., n$, where U is concave, continuous, monotone and nonsatiated (Afriat [Afr67], Varian [Var83]). Hence we may pose the following question for any two unobserved consumption bundles $\bar{x}$ and $\hat{x}$: Will $\bar{x}$ be revealed preferred to $\hat{x}$ for every solution of the Afriat inequalities? An equivalent for-mulation is the counterfactual proposition: $\bar{x}$ is not revealed preferred to $\hat{x}$ for some price vector p and some utility function U, which is a solution of the Afriat inequalities on $(p_1, x_1), (p_2, x_2), \ldots, (p_n.x_n)$.

This proposition can be expressed in terms of the solution set for the following family of polynomial inequalities: The Afriat inequalities for the augmented data set $(p_1, x_1), (p_2, x_2), ..., (p_n, x_n)$, $(p, \hat{x})$ and the inequality $p \cdot \bar{x} > p \cdot \hat{x}$, where p is unobserved. If these inequalities are solvable, then the stated counterfactual proposition is true. If not, then the answer to our original question is yes. Notice that n of the Afriat inequalities are quadratic multi-variate polynomials in the unobservables, i.e., the product of the marginal utility of income at $\hat{x}$ and the unknown price vector p.

We extend the analysis of Brown and Matzkin [BM96], where the Wal-rasian equilibrium inequalities are derived, to encompass the computation of counterfactual equilibria in Walrasian economies.

2 Economic Models

Brown and Matzkin [BM96] characterized the Walrasian model of compet-itive market economies for data sets consisting of a finite number of obser-vations on market prices, income distributions and aggregate demand. The Walrasian equilibrium inequalities, as they are called here, are defined by the Afriat inequalities for individual demand and budget constraints for each con-sumer; the Afriat inequalities for profit maximization over a convex aggregate technology; and the aggregation conditions that observed aggregate demand is the sum of unobserved individual demands. The Brown–Matzkin theorem states that market data is consistent with the Walrasian model if and only if the Walrasian equilibrium inequalities are solvable for the unobserved utility levels, marginal utilities of income and individual demands. Since individual demands are assumed to be unobservable, the Afriat inequalities for each consumer are quadratic multivariate polynomials in the unobservables, i.e., the product of the marginal utilities of income and individual demands.[3]

We consider an economy with L commodities and T consumers. Each agent has $\mathbb{R}_+^L$ as her consumption set. We restrict attention to strictly positive market prices $S = \{p \in \mathbb{R}_{++}^L : \sum_{i=1}^L p_i = 1\}$. The Walrasian model assumes that consumers have utility functions $u_t : \mathbb{R}_+^L \to \mathbb{R}$, income I_t and that aggregate demand $\bar{x} = \sum_{t=1}^T x_t$, where

$$u_t(x_t) = \max_{p \cdot x \leq I_t, x \geq 0} u_t(x).$$

Suppose we observe a finite number N of profiles of incomes of consumers $\{I_t^r\}_{t=1}^T$, market prices p^r and aggregate demand $\bar{x}^r$, where $r = 1, 2, ..., N$, but we do not observe the utility functions or demands of individual consumers. When are these data consistent with the Walrasian model of aggregate demand? The answer to this question is given by the following theorems of Brown and Matzkin [BM96].

Theorem 1 (Theorem 2, Brown–Matzkin). *There exist nonsatiated, continuous, strictly concave, monotone utility functions $\{u_t\}_{t=1}^T$ and $\{x_t^r\}_{t=1}^T$, such that $u_t(x_t^r) = \max_{p^r \cdot x \leq I_t^r} u_t(x)$ and $\sum_{t=1}^T x_t^r = \bar{x}^r$, where $r = 1, 2, ..., N$, if and only if $\exists \{\hat{u}_t^r\}, \{\lambda_t^r\}$ and $\{x_t^r\}$ for $r = 1, ..., N; t = 1, ..., T$ such that*

$$\hat{u}_t^r < \hat{u}_t^s + \lambda_t^s p^s \cdot (x_t^r - x_t^s) \quad (r \neq s = 1, ..., N; \ t = 1, ..., T) \tag{1}$$

$$\lambda_t^r > 0, \quad (t = 1, 2, ... T; \ r = 1, 2, ... N) \tag{2}$$

$$\hat{u}_t^r > 0 \text{ and } x_t^r \geq 0 \quad (r = 1, ..., N; \ t = 1, ..., T) \tag{3}$$

$$p^r \cdot x_t^r = I_t^r \quad (r = 1, ..., N; \ t = 1, ..., T) \tag{4}$$

$$\sum_{t=1}^T x_t^r = \bar{x}^r \quad (r = 1, ..., N) \tag{5}$$

Equations (1), (2) and (3) constitute the strict Afriat inequalities; (4) defines the budget constraints for each consumer; and (5) is the aggregation condition that observed aggregate demand is the sum of unobserved individual consumer demand. This family of inequalities is called here the (strict) Walrasian equilibrium inequalities. [4] The observable variables in this system of inequalities are the I_t^r, p^r and $\bar{x}^r$, hence this is a nonlinear family of multivariate polynomial inequalities in unobservable utility levels $\hat{u}_t^r$, marginal utilities of income λ_t^r and individual consumer demands x_t^r.

---

[3] The Afriat inequalities for competitive profit maximizing firms are linear given market data—see Varian [Var84]. Hence we limit our discussion to the nonlinear Afriat inequalities for consumers.

[4] Brown and Matzkin call them the equilibrium inequalities, but there are other plausible notions of equilibrium in market economies.

The case of homothetic utilities is characterized by the following theorem of Brown and Matzkin [BM96].

Theorem 2 (Theorem 4, Brown–Matzkin). *There exist nonsatiated, continuous, strictly concave homothetic monotone utility functions* $\{u_t\}_{t=1}^T$ *and* $\{x_t^r\}_{t=1}^T$ *such that* $u_t(x_t^r) = \max_{p^r \cdot x \leq I_t^r} u_t(x)$ *and* $\sum_{t=1}^T x_t^r = \bar{x}_t^r$, *where* $r = 1, 2, ..., N$ *if and only if* $\exists \{\hat{u}_t^r\}$ *and* $\{x_t^r\}$ *for* $r = 1, ..., N$; $t = 1, ..., T$ *such that*

$$\hat{u}_t^r < \hat{u}_t^s \frac{p^s \cdot x_t^r}{p^s \cdot x_t^s} \quad (r \neq s = 1, ..., N; \ t = 1, ..., T) \tag{6}$$

$$\hat{u}_t^r > 0 \ and \ x_t^r \geq 0 \quad (r = 1, ..., N; \ t = 1, ..., T) \tag{7}$$

$$p^r \cdot x_t^r = I_t^r \quad (r = 1, ..., N; \ t = 1, ..., T) \tag{8}$$

$$\sum_{t=1}^T x_t^r = \bar{x}^r \quad (r = 1, ..., N) \tag{9}$$

Equations (6) and (7) constitute the strict Afriat inequalities for homothetic utility functions.

The Brown–Matzkin analysis extends to production economies, where firms are price-taking profit maximizers. See Varian [Var84] for the Afriat inequalities characterizing the behavior of firms in the Walrasian model of a market economy.

3 Algorithms

An algorithm for solving the Walrasian equilibrium inequalities constitutes a specification test for multiple calibration of numerical general equilibrium models, i.e., the market data is consistent with the Walrasian model if and only if the Walrasian equilibrium inequalities are solvable.

In multiple calibration, two or more years of market data together with empirical studies on demand and production functions and the general equilibrium restrictions are used to specify numerical general equilibrium models. The maintained assumption is that the market data in each year is consistent with the Walrasian model of market economies. This assumption which is crucial to the calibration approach is never tested, as noted in Dawkins et al.

The assumption of Walrasian equilibrium in the observed markets is testable, under a variety of assumptions on consumer's tastes, using the necessary and sufficient conditions stated in Theorems 1 and 2 and the market data available in multiple calibration. In particular, Theorem 2 can be used as a specification test for the numerical general equilibrium models discussed in Shoven and Whalley [SW92], where it is typically assumed that utility functions are homothetic.

If we observe all the exogenous and endogenous variables, as assumed by Shoven and Whalley, then the specification test is implemented by solving the linear program, defined by (1), (2), (3), (4) and (5) for utility levels and marginal utilities of income or in the homothetic case, solving the linear program defined by (6), (7), (8), and (9) for utility levels.

If individual demands for goods and factors are not observed then the specification test is implemented using the deterministic algorithm presented here.

Following Varian, we can extrapolate from the observed market data available in multiple calibration to unobserved market configurations. We simply augment the equilibrium inequalities defined by the observed data with additional multivariate polynomial inequalities characterizing possible but unobserved market configurations of utility levels, marginal utilities of income, individual demands, aggregate demands, income distributions and equilibrium prices. Counterfactual equilibria are defined as solutions to this augmented family of equilibrium inequalities.

In general, the Afriat inequalities in this system will be cubic multivariate polynomials because they involve the product of unobserved marginal utilities of income, the unobserved equilibrium prices and unobserved individual demands. If the observations include the market prices then the Afriat inequalities are only quadratic multivariate polynomials in the product of the unobserved marginal utility of income and individual demand. We now present the determinisitic algorithm for the quadratic case.

3.1 Deterministic Algorithm

The algorithm is based on the simple intuition that if one knows the order of the utility levels over the observations for each consumer, then all we have to do is to solve a linear program. We will enumerate all possible orders and solve a linear program for each order. An important point is that the number of orders is $(N!)^T$, where, N is the number of observations and T the number of agents. Hence the algorithm will run in time bounded by a function which is polynomial in the number of commodities and exponential only in N and T. In situations involving a large number of commodities and a small N, T, this is very efficient. Note that trade between countries observed over a small number of observations is an example.

Consider the strict Afriat inequalities (1), (2) and (3) of Theorem 1.

Since the set of $\hat{u}_t^r$ for which there is a solution to these inqualities is an open set, we are free to add the condition: No two $\hat{u}_t^r$ are equal. This technical condition will ensure a sort of "non-degenracy." Under this condition, using the concavity of the utility function, it can be shown that (1) and (2) are equivalent to

$$\hat{u}_t^r > \hat{u}_t^s \Rightarrow p^s \cdot x_t^r > p^s x_t^s \quad (r \neq s = 1, ..., N;\ t = 1, ..., T). \tag{10}$$

The system (10) is not a nice system of multivariate polynomial inequalities. But now, consider a fixed consumer t. Suppose we fix the order of the $\{\hat{u}_t^r : r = 1, 2, \ldots N\}$. Then in fact, we will see that the set of feasible consumption vectors for that consumer is a polyhedron. Indeed, let σ be a permuation of $\{1, 2 \ldots N\}$. σ will define the order of the $\{\hat{u}_t^r : r = 1, 2, \ldots, N\}$; i.e., we will have

$$\hat{u}_t^{\sigma(1)} < \hat{u}_t^{\sigma(2)} < \cdots < \hat{u}_t^{\sigma(N)}. \tag{11}$$

Then define $P(\sigma, I_t)$ to be the set of $x = (x_t^1, x_t^2, \ldots, x_t^N)$ satsifying the following:

$$x_t^r \in \mathbb{R}_{++}^L \tag{12}$$

$$p^{\sigma(s)} \cdot x_t^{\sigma(r)} > p^{\sigma(s)} x_t^{\sigma(s)} \quad N \geq r > s \geq 1 \tag{13}$$

$$p^r \cdot x_t^r = I_t^r \quad 1 \leq r \leq N \tag{14}$$

Lemma 1. $P(\sigma, I_t)$ *is precisely the set of* $(x_t^1, x_t^2, \ldots, x_t^N)$ *for which there exist* λ_t^r, $\hat{u}_t^r$ *satisfying* (11) *so that* $\hat{u}_t^r$, λ_t^r, x_t^r *together satisfy* (1), (2), (3) *and* (4).

Now, fix a set of T permutations—$\sigma_1, \sigma_2, \ldots, \sigma_T$, one for each consumer. We can then write a linear system of inequalities for $P(\sigma_t, I_t)$ for $t = 1, 2, \ldots, T$ and the consumption total (5). If this system is feasible, then there is a feasible solution with the utilities in the given order. Clearly, if all such $(N!)^T$ systems are infeasible, then there is no rationalization.

Remark. If for a particular consumer t,

$$\frac{p^1}{I_t^1} \geq \frac{p^2}{I_t^2} \geq \cdots \geq \frac{p^N}{I_t^N},$$

then, by Lemma 1 of Brown–Shannon [BS00], we may assume that the u_t^r are non-decreasing. For such a consumer, only one order needs to be considered. Note that the above condition says that in a sense, the consumer's income outpaces inflation on each and every good.

A more challenging problem is the computation of counterfactual equilibria. Fortunately, a common restriction in applied general equilibrium analysis is the assumption that consumers are maximizing homothetic utility functions subject to their budget constraints and firms have homothetic production functions. A discussion of the Afriat inequalities for cost minimization and profit maximization for firms with homothetic production functions can be found in Varian [Var84]. Afriat [Afr81] and subsequently Varian [Var83] derived a family of inequalities in terms of utility levels, market prices and incomes that characterize consumer's demands if utility functions are homothetic. We shall refer to these inequalities as the homothetic Afriat inequalities.

Following Shoven and Whalley [SW92, p. 107], we assume that we observe all the exogenous and endogenous market variables in the benchmark equilibrium data sets, used in the calibration exercise. As an example, suppose there is only one benchmark data set, then the (strict) homothetic Afriat inequalities for each consumer are of the form:[5]

$$U^1 < \lambda^2 p^2 \cdot x^1 \text{ and } U^2 < \lambda^1 p^1 \cdot x^2$$
$$U^1 = \lambda^1 I^1 \qquad\qquad U^2 = \lambda^2 I^2$$

where we observe p^1, x^1 and I^1. Given λ^1 and λ^2 we again have a linear system of inequalities in the unobserved U^1, U^2, x^2, p^2 and I^2. A similar set of inequalities can be derived for cost minimizing or profit maximizing firms with production functions that are homogenous of degree one.

3.2 VC algorithm

For the general case (with no assumptions on homotheticity or knowledge of all price vectors) , we propose the "VC algorithm"; here we give a brief description of this with details to follow. First, let y be a vector consisting of marginal utilities of income and utility levels of each consumer in each observation (so $y \in \mathbb{R}^k$, where $k = 2NT$). Let x be a vector comprising the consumption vectors for each consumer in each observation (so $x \in \mathbb{R}^n$, where $n = NTL$). Finally, let z be a vector comprising the unknown price vector in the case of the counterfactual proposition. Then our inequalities are multivariate polynomials in the variables (x, y, z); but they are multi-linear in the x, y, z. We would like to find a solution (x, y, z) in case one exists. Suppose now that the proportion of all (x, y, z) which are solutions is some $\delta > 0$. A simple algorithm for finding a solution would be to pick at random many (x, y, z) triples and check if any one of these solve the inequalities. The number of trials we need is of the form $1/\delta$, In general, δ may be very small (even if positive) and we have to do many trials. Instead consider another parameter—ε defined by

$$\varepsilon = \max{}_y \text{ (proportion of } (x, z) \text{ such that } (x, y, z) \text{ solve the inequalities.)}$$

Clearly, ε is always at least δ, but it is typically much larger. Using the concept of Vapnik–Čherovenkis (VC) dimension, we will give an algorithm which makes $O(d/\varepsilon)$ trials, where d is the VC dimension of the VC class of sets defined by our inequalities. We elaborate on this later, but remark here that d is $O(NT(\log NTL))$. Thus when at least one y admits a not too small proportion of (x, z) as solutions, this algorithm will be efficient.

To develop the algorithm, we construct an ε-net—a small set S of (x, z) such that with high probability, every set in the (x, z) space containing at least ε proportion of the (x, z) 's has an element of S. This ε-net is derived from the VC-property of the Walrasian equilibrium inequalities. That is, Laskowski [Las92b] has shown that any finite family of multivariate polynomial inequalities in unknowns (x, y, z), defines a VC-class of sets. As is well known in computational geometry, VC-classes imply the existence of "small" ε-nets.

---

[5] Here we assume utility functions are homogenous of degree one.

In our setting, at each point in the ε-net, the Walrasian inequalities define a linear program in the y 's , since the Walrasian equilibrium inequalities are a system of multilinear polynomial inequalities. One of these linear programs will yield a solution to our system. Of course, ε as defined above is not known in advance. We just try first 1 as a value for ε and if this fails, we half it and so on until we either have success or the value tried is small enough, whence we may stop and assert with high confidence that for every y, the proportion of (x, z) which solve the inequalities is very small. (This includes the case of an empty set of solutions.) This algorithm is polynomial in the number of variables and $1/\varepsilon$. Now the details.

A collection $\mathcal{C}$ of subsets of some space χ picks out a certain subset E of a finite set $\{x_1, x_2, ..., x_n\} \subset \chi$ if $E = \{x_1, ..., x_n\} \cap A$ for some $A \in \mathcal{C}$. $\mathcal{C}$ is said to shatter $\{x_1, ..., x_n\}$ is $\mathcal{C}$ picks out each of its 2^n subsets. The VC-dimension of $\mathcal{C}$, denoted $V(\mathcal{C})$ is the smallest n for which no set of size n is shattered by $\mathcal{C}$. A collection $\mathcal{C}$ of measurable sets is called a VC-class if its dimension, $V(\mathcal{C})$, is finite.

Let y be a vector consisting of marginal utilities of income and utility levels of each consumer in each observation (so $y \in \mathbb{R}^k$, where $k = 2NT$). Let x be a vector comprising the consumption vectors for each consumer in each observation (so $x \in \mathbb{R}^n$, where $n = NTL$). Finally, let z be a vector comprising the unknown price vector in the case of the counterfactual proposition (so $z \in \mathbb{R}^L$). The Walrasian equilibrium inequalities (1) through (5) define a Boolean formula $\Phi(x, y, z)$ containing $s \in O(N^2 T + NTL)$ atomic predicates where each predicate is a polynomial equality or inequality over $n + k + L$ variables—x, y, z. Here Φ is simply the conjunction of the inequalities (1) through (5). For any fixed y, let $F_y \subseteq \mathbb{R}^{n+L}$ be the set of (x, z) such that $\Phi(x, y, z)$ is satisfied then this is a VC-class of sets by Laskowski's Theorem. Let $\varepsilon = \max_y$ (proportion of (x, z) such that (x, y, z) solve the inequalities.) Note that if the set of solutions has measure 0 (or in particular is empty), then $\varepsilon = 0$.

Theorem 3 (Goldberg–Jerrum [GJ95]). *The family $\{F_y : y \in R^k\}$ has VC-dimension at most $2k \log_2(100s)$.*

The next proposition is a result from Blumer et al. [BEHW89].

Theorem 4. *If F is a VC-class of subsets of X with VC-dimension d, and ε, δ are positive reals and $m \geq (8d/\varepsilon\delta^2) \log_2(13/\varepsilon)$, then a random[6] subset $\{x_1, x_2, ..., x_m\}$ of X (called a ε-net) satisfies the following with probabilty at least $1 - \delta$:*

$$S \cap \{x_1, ..., x_m\} \neq \phi \quad \forall S \in \mathcal{F} \text{ with } \mu(S) \geq \varepsilon.$$

For fixed $\bar{x}$, $\bar{z}$, the Walrasian equilibrium inequalities define a linear program over $y \in \mathbb{R}_+^k$. Hence each point in the ε-net defines a linear program in

---

[6] Picked in independent, uniform trials.

the y's. Now, the random algorithm is clear: we pick a random set of $m(x, z)$ pairs, where ε is defined above and solve a linear program for each. For the correct ε, we will succeed with high probability in finding a solution. We do not know the ε a priori, but we use a standard trick of starting with $\varepsilon = 1$ and halving it each time until we have success or we stop with a small upper bound on ε.

A preliminary version of the VC-algorithm was introduced in Brown and Kannan [BK03]. But the bound here on the VC dimension using Theorem 5 and hence the running time are substantially better than those derived in that report.

Acknowledgments

This revision of [BK03] has benefited from our reading of Kubler's [Kub07]. We thank him, Glena Ames and Louise Danishevsky for their assistance.

Brown, D.J., Kannan, R.: Two algorithms for solving the Walrasian inequalities. Cowles Foundation Discussion Paper No. 1508R, Yale University (2006). Reprinted by permission of the Cowles Foundation.

Is Intertemporal Choice Theory Testable?

Felix Kubler

University of Pennsylvania, Philadelphia, PA 19104-6297 kubler@sas.upenn.edu

Summary. Kreps–Porteus preferences constitute a widely used alternative to time separability. We show in this paper that with these preferences utility maximization does not impose any observable restrictions on a household's savings decisions or on choices in good markets over time. The additional assumption of a weakly separable aggregator is needed to ensure that the assumption of utility maximization restricts intertemporal choices. Under this assumption, choices in spot markets are characterized by a strong axiom of revealed preferences (SSARP).

Under uncertainty Kreps–Porteus preferences impose observable restrictions on portfolio choice if one observes the last period of an individual's planning horizon. Otherwise there are no restrictions.

Key words: Intertemporal choice, Non-parametric restrictions

1 Introduction

There is a large literature on testing individual demand data for consistency with utility maximization (see, e.g., Afriat [Afr67], Varian [Var82], Chiappori and Rochet [CR87]). In this literature, it is assumed that one observes how an individual's choices vary as prices and his income vary. However, data of this sort can only be obtained through experiments. If one actually records an individual's actions in markets over time, these classical tests of demand theory might be useless because they neglect the fact that an agent's choices today may be affected by his choice set tomorrow or his savings from previous periods. Tests of demand theory which use market data must be tests of intertemporal choice models. If one assumes that all agents maximize time-separable and time-invariant utility and if one only observes their choices in spot markets (i.e., saving decisions or incomes are unobservable) the analysis in Chiappori and Rochet [CR87] remains valid and a strong version of the strong axiom of revealed preferences (SSARP, see Chiappori and Rochet [CR87]) is necessary and sufficient for data to be consistent with utility maximization. However, time separability is a very strong restriction on preferences which holds only if one assumes that preference orderings on consumption streams from $t = 1, ..., s$ are independent of an agent's expectations on his consumptions for the periods from $s + 1$ onwards.

While it seems intuitively reasonable to argue that history independence and time consistency together with some form of stationarity is enough to ensure that an agent's choice behavior is restricted by the assumption of utility maximization, we show that this intuition is wrong and that the assumption of Kreps–Porteus preferences [KP78] does not impose any restriction on observed choices. It follows from our analysis that such widely used concepts as time consistency or pay-off history independence are not testable if one does not use experimental data but is confined to data on individual behavior in markets.

Uncertainty adds an additional dimension to the agent's choice problem. Risk aversion will generally impose restrictions on portfolio selection when continuation utilities are identical across all possible next period states. The question then arises to what extend these restrictions are observable. If one observes the last period of an individual's planning horizon, these restrictions are reflected in the individual's portfolio holdings coming into this last period. However, if we assume that the last period is not observable, the assumption of Kreps–Porteus preferences imposes no restrictions on portfolio selection even when all off sample path choices as well as all probabilities are observable.

These negative results raise the questions under which conditions utility maximization does impose restrictions on intertemporal choices. We derive a sufficient (additional) condition on the aggregator function, which ensures that the model is testable. If the aggregator function is weakly separable then choices on spot markets must satisfy SSARP. If asset prices are unobservable, SSARP is also sufficient for the choices to be rationalizable by a time-separable utility function and the two specifications are therefore observationally equivalent.

We develop our arguments for a finite horizon choice problem.

Without stationarity assumptions, as long as the number of observed choices is finite, one cannot refute the conjecture that the agent maximizes a Kreps–Porteus style utility function over an infinite horizon consumption program. However, it seems natural to impose a Markov structure on the infinite horizon problem and to confine attention to recursive utility of the Epstein–Zin type [EZ89]. An extension of these results to the infinite horizon problem is subject to future research.

The paper is organized as follows. In Section 2 we introduce the model and some notation. Section 3 proves the main result and discusses its implications for a finite horizon choice problems, both under certainty and under uncertainty.

2 The Model

We consider an individual's choice problem over $\bar{T}+1$ periods, $t = 0, ..., \bar{T}$ with uncertainty resolving each period. We take as given an event tree Ξ with nodes $\xi \in \Xi$. Let ξ_0 be the root node, i.e., the unique node without a predecessor. For all other nodes, let ξ_- be the unique predecessor of node ξ. For all nodes $\xi \in \Xi$, let $\mathcal{J}(\xi)$ be the set of its immediate successors. Nodes without successors, i.e., $\mathcal{J}(\xi)$ is empty, are called terminal nodes. Finally, we collect all nodes which are possible at some period t in a set $\mathcal{N}_t$ and we denote by M the total number of nodes in the event tree. We assume that M is finite. For simplicity, we assume that there are no terminal nodes in any $\mathcal{N}_t$ for $t < T$.

At each node ξ there are J short-lived assets with asset j paying $d_j(\zeta) \in \mathbb{R}$ at all nodes $\zeta \in \mathcal{J}(\xi)$, its price being denoted by $q_j(\xi)$.

At each node $\xi \in \Xi$, the individual receives an exogenous income $I(\xi) \in \mathbb{R}_+$ (either from selling endowments or from transfers) and he is active in spot and asset markets. He faces prices $p(\xi) \in \mathbb{R}^L_{++}$ and chooses a consumption bundle $c(\xi) \in \mathbb{R}^L_+$.

The agent's consumption decisions must be supported by portfolio choices $(\theta(\xi))_{\xi \in \Xi}$, $\theta(\xi) \in \mathbb{R}^J$. All consumptions and portfolio choices $(c(\xi), \theta(\xi))_{\xi \in \Xi}$ must lie in the individual's budget set which we define as

$$B\left((p(\xi), q(\xi), d(\xi), I(\xi))_{\xi \in \Xi}\right) = \{(c(\xi), \theta(\xi))_{\xi \in \Xi} : p(\xi)c(\xi) + q(\xi)\theta(\xi) \leq I(\xi)$$
$$+ \theta(\xi_-)d(\xi), c(\xi) \geq 0 \text{ for all } \xi \in \Xi\}$$

where we normalize $\theta(\xi_{0-}) := 0$.

The agent attaches a positive probability to each node. Given a node $\zeta \in \Xi$ and a direct successor $\xi \in \mathcal{J}(\xi)$, we denote by $\mu(\xi)$ the (unconditional) probability of node ξ and by $\mu(\xi|\zeta)$ the conditional probability of ξ given ζ.

We say that an agent's utility function $u : \mathbb{R}^{LM}_+ \to \mathbb{R}$ is of the Kreps–Porteus type if $u((c(\xi))_{\xi \in \Xi}) = v_{\xi 0}$, where v_ξ, utility at node is ξ recursively defined by

$$v_\xi(c(\xi)) = W(c(\xi), \mu(\xi))$$

with

$$\mu(\xi) = \sum_{\zeta \in \mathcal{J}(\xi)} \pi(\zeta|\xi)v_\zeta(c(\zeta)) \quad \text{for all non-terminal } \xi.$$

We will assume throughout that the aggregator $W : \mathbb{R}^L_+ \times \mathbb{R}_+ \to \mathbb{R}_+$ is twice continuously differentiable, strictly increasing and strictly concave. We normalize $W(0,0) = 0$ and hence impose $\mu(\xi) = 0$ for terminal nodes ξ.

We also impose two regularity conditions on the aggregator which are often needed to extend the preference specification to infinite horizon problems (see, e.g., Koopmans [Koo60] or Epstein and Zin [EZ89]).

LS1: The function $W(\cdot, \cdot)$ is bounded, i.e.,

$$\sup_{x \in \mathbb{R}^L_+, y \geq 0} W(x, y) < \infty.$$

LS2: The second partial derivative of $W(\cdot,\cdot)$ is bounded above by one, i.e.,

$$\partial_y W(x,y) < 1 \text{ for all } x \in \mathbb{R}_L$$

In a slight abuse of notation we will refer to utility functions which satisfy all of the above assumptions as 'Kreps–Porteus utility'.

2.1 Observations

In order to make our main argument, we assume that we observe choices, prices and incomes at periods $t = 0, ..., T \leq \bar{T}$. Under uncertainty, we assume that we observe these variables and all dividends all nodes $\xi \in \mathcal{N}_t$, $t = 0, ..., T$ as well as all relevant probabilities (which might be known when we assume objective laws of motion). In order to present our main argument as strong as possible, we assume that all last period continuation utilities $\mu(\xi)$ for all $\xi \in \mathcal{N}_T$ are known (which might justified because they are all zero and it is the last period of the individual's planning horizon, i.e., $T = \bar{T}$). When we discuss our result below, we will assess how realistic these assumptions are. We define $\Omega = \cup_{t=1}^T \mathcal{N}_t$ be the set of all observable nodes in the event tree. An extended observation is then given by

$$\mathcal{O} = ((c(\xi), \theta(\xi), d(\xi), q(\xi), p(\xi), \pi(\xi))_{\xi \in \Omega}, (\mu(\xi))_{\xi \in \mathcal{N}_T}).$$

The question is whether there are restrictions on this observations imposed by the assumption of Kreps–Porteus utility. It is important to note that if one does not observe an agent's choices over his entire planning horizon (i.e., if $\bar{T} > T$) one is free to choose choices as well as prices, dividends and incomes at all nodes which are not in Ω. We therefore have the following definition.

Definition 1. An extended observation

$$\mathcal{O} = ((c(\xi), \theta(\xi), d(\xi), q(\xi), p(\xi), \pi(\xi))_{\xi \in \Omega}, (\mu(\xi))_{\xi \in \mathcal{N}_T})$$

is said to be rationalizable by Kreps–Porteus utility if there exist $c(\xi)$, $\theta(\xi)$, $p(\xi)$, $q(\xi)$, $I(\xi)$, $d(\xi)$ for all $\xi \in \Xi$, $\xi \notin \Omega$ and if there exists a Kreps–Porteus utility function $u(\cdot)$ which is consistent with the probabilities $(\pi(\xi))_{\xi \in \Omega}$ and the last period continuation utilities $\mu(\xi)$, $\xi \in \mathcal{N}_T$ such that

$$(c(\xi), \theta(\xi))_{\xi \in \Xi} \in \arg\max_{c \in \mathbb{R}_+^{LM}, \theta \in \mathbb{R}^{JM}} u(c)$$

such that

$$(c(\xi), \theta(\xi)_{\xi \in \Xi} \in \mathcal{B}((p(\xi), q(\xi), (d(\xi), (I(\xi))_{\xi \in \Xi}).$$

It is well known that the absence of arbitrage is a necessary condition for the agent's choice problem to have a finite solution

Definition 2. Prices and dividends $(p(\xi), q(\xi), d(\xi))_{\xi \in \Xi}$ preclude arbitrage if there is no trading strategy $(\theta(\xi))_{\xi \in \Xi}$ with $\theta_{0-} = 0$ such that if we define

$$D^{\theta}(\xi) = \theta(\xi_-)d(\xi) - \theta(\xi)q(\xi)$$
$$D^{\theta}(\xi) \geq 0 \text{ for all } \xi \in \Xi \text{ and } D^{\theta} \neq 0.$$

We will assume throughout that the observed prices preclude arbitrage and that the observed choices lie in the agent's budget set. We also assume that we never observe zero consumption, i.e., that $c(\xi) \neq 0$ for all $\xi \in \Omega$ (although consumption of a given commodity might sometimes be zero, it cannot be the case that the agent chooses to consume nothing at all) and that the agent does not trade assets in the last period of his planning horizon, $\theta(\xi) = 0$ for all $\xi \in \mathcal{N}_{\bar{T}}$. Finally, we assume that $\mu(\xi) = 0$ for terminal nodes $\xi \in \mathcal{N}_{\bar{T}}$. These restrictions on observed choices are trivial restrictions and follow directly from monotonicity.

3 Observable Restrictions

For our non-parametric analysis we need to derive Afriat inequalities [Afr67]. These non-linear inequalities completely characterize choices which are consistent with the maximization of a Kreps–Porteus utility function.

Lemma 1. *An extended observation*

$$\mathcal{O} = ((c(\xi), \theta(\xi), d(\xi), q(\xi), p(\xi), \pi(\xi))_{\xi \in \Omega}, (\mu(\xi))_{\xi \in \mathcal{N}_T})$$

with all $c(\xi) \in \mathbb{R}_{++}^L$ is rationalizable by a Kreps–Porteus utility function if and only if there exist positive numbers $\lambda(\xi)$, $\eta(\xi)$, $\gamma(\xi)$, $W(\xi))_{\xi \in \Omega}$ with $\gamma(\xi) < 1$ for all $\xi \in \Omega$, with

$$\eta(\xi) = \sum_{\zeta \in \mathcal{J}(\xi)} \pi(\zeta|\xi)W(\zeta)$$

for all $\xi \in \mathcal{N}_t$, $t < T$ such that

- *for all $\xi \in \mathcal{N}_t$, $t < T$,*

$$\lambda(\xi)q_j(\xi) = \gamma(\xi) \sum_{\zeta \in \mathcal{J}(\xi)} \pi(\zeta|\xi)\lambda(\zeta)d_j(\zeta) \text{ for } j = 1, ..., J \qquad (U1)$$

- *for all $\xi, \zeta \in \Omega$,*

$$W(\xi) - W(\zeta) \leq \lambda(\zeta)p(\zeta)(c(\xi) - c(\zeta)) + \gamma(\zeta)(\eta(\xi) - \eta(\zeta)) \qquad (U2)$$

the inequality holds strict if $c(\xi) \neq c(\zeta)$ or if $\eta(\xi) \neq \eta(\zeta)$.

If for some node $\xi \in \Omega$, the observed consumption, $c(\xi)$, lies on the boundary of $\mathbb{R}_+^L$ the conditions remain sufficient but are no longer necessary.

Proof. For the necessity part, consider the agent's first-order condition (which are necessary and sufficient for optimality of interior choices):

At any node $\xi \in \Xi$,

$$\partial_c W(c(\xi), \mu(\xi)) - \lambda(\xi) p(\xi) = 0$$

and

$$\eta(\xi) q_j(\xi) = \sum_{\zeta \in \mathcal{J}(\xi)} \eta(\zeta) d_j(\zeta) \quad \text{for } j = 1, ..., J \text{ and for all non-terminal } \xi \in \Xi$$

where $\lambda(\xi_0) = \eta(\xi_0)$ and where for all $\xi \neq \xi_0 \in \Xi$,

$$\lambda(\xi) = \lambda(\xi_-) \frac{\eta(\xi)}{\pi(\xi|\xi_-)\partial_\mu W(c(\xi_-), \mu(\xi_-))}.$$

Defining $\gamma(\xi) = \partial_\mu W(c(\xi), \mu(\xi))$, these first-order condition, together with the assumption that $W(\cdot, \cdot)$ is concave and the usual characterization of concave functions proves necessity: (1) stems from the second set of first-order conditions and inequality (1) characterizes strict concavity of $W(\cdot)$, where the first optimality condition is used to substitute for $\partial_c W$. The assumption that $\gamma(\xi) < 1$ for all $\xi \in \Omega$ follows from condition LS2.

For the sufficiency part, assume that the unknown numbers exist and satisfy the inequalities. We can then construct a piecewise linear aggregator function following Varian [Var82].

Define

$$W(c, \mu) = \min_{\xi \in \Omega} \left\{ U(\xi) + \begin{pmatrix} \lambda(\xi) \\ \gamma(\xi) \end{pmatrix} \left[\begin{pmatrix} p(\xi)c \\ \mu \end{pmatrix} - \begin{pmatrix} p(\xi)c(\xi) \\ \eta(\xi) \end{pmatrix} \right] \right\}.$$

The resulting function is clearly concave and strictly increasing and the function rationalizes the observation $\mathcal{O}$. Furthermore, the approach in Chiappori and Rochet [CR87] can be used to construct a strictly concave and smooth aggregator function. Their argument goes through without any modification.

Since $\gamma(\xi) < 1$ for all ξ it follows immediately that LS2 must hold. LS1 follows from the fact that all constructed numbers are finite.

When $T < \bar{T}$, we can construct future dividends, prices and consumptions such that they are consistent with period T portfolio holdings and period T continuation utilities. The key is to observe that for all possible observed continuation utilities $\mu(\xi)$, $\xi \in \mathcal{N}_T$ and for all last period portfolios $\theta(\xi)$, $\xi \in \mathcal{N}_T$ there will exist unobserved next period dividends to rationalize them. $\square$

We now use this characterization to show that the assumption of Kreps–Porteus utility is in general not testable using only market data since it imposes very few restrictions on observed choices.

Theorem 1. *Any possible extended observation $\mathcal{O}$ for which $\mu(\xi) \neq \mu(\zeta)$ for all nodes $\xi \neq \zeta \in \mathcal{N}_T$ can be rationalized by a Kreps–Porteus utility function.*

The following lemma is crucial for the proof of the theorem. While the lemma appears simple its proof turns out to be quite tedious.

Lemma 2. *For any finite event tree Ω, probabilities $(\pi(\xi))_{\xi \in \Omega}$ and positive numbers $\eta(\xi)$ for all terminal $\xi \in \Omega$ with $\eta(\xi) \neq \eta(\zeta)$ for all $\xi \neq \zeta$ there exist a $\bar{\gamma} > 0$, $(W(\xi), \gamma(\xi))_{\xi \in \Omega}$, $1 > \gamma(\xi) \geq \bar{\gamma}$ and $W(\xi) > 0$ for all $\xi \in \Omega$ as well as a number $\delta > 0$ such that*

$$W(\zeta) - W(\xi) + \gamma(\zeta)(\eta(\xi) - \eta(\zeta)) > \delta \quad \text{for all } \zeta, \xi \in \Omega \tag{1}$$

with

$$\eta(\xi) = \sum_{\zeta \in \mathcal{J}(\xi)} \pi(\zeta|\xi) W(\xi) \quad \text{for all non-terminal } \xi \in \Omega.$$

Proof. We construct these number recursively. Let T denote the number of periods in Ω, let m denote the number of nodes in Ω and fix some $\varepsilon < 1/(m+2)$. Fix δ to ensure that $0 < \delta < \varepsilon$.

If $n_t = \#\mathcal{N}_t$ denotes the number of nodes at period t, we can define a function $\xi_t(i)$ by $\eta(\xi_t(1)) < \eta(\xi_t(2)) < \cdots < \eta(\xi_t(n_t))$. Since by assumption $\eta(\xi) \neq \eta(\zeta)$ for all $\xi, \zeta \in \mathcal{N}_T$, this function exists for $t = T$.

For T we can choose the associated $\gamma(\xi)$ such that

$$1 = \varepsilon = \gamma(\xi_T(1)) = \gamma(\xi_T(2)) + \varepsilon = \cdots = \gamma(\xi_T(n_T)) + (n_t - 1)\varepsilon.$$

Now choose $W(\xi_T(1)) > \eta(\xi_T(n_T))$ and define for $i = 2, ..., n_T$

$$W(\xi_T(i)) = W(\xi_T(i-1)) + \gamma(\xi_T(i-1))(\eta(\xi_T(i)) - \eta(\xi_T(i=1))) - \delta.$$

Given $(W(\xi), \gamma(\xi))_{\xi \in \mathcal{N}_t}$ we can construct $(\eta(\xi), W(\xi), \gamma(\xi))$ for $\xi \in \mathcal{N}_{t-1}$ as follows: For all $\xi \in \mathcal{N}_{t-1}$, compute the new $\eta(\xi) = \sum_{\zeta \in \mathcal{J}(\xi)} \pi(\zeta|\xi) W(\zeta)$. One can choose δ to ensure that $\eta(\xi) \neq \eta(\zeta)$ for all $\xi, \zeta \in \mathcal{N}_{t-1}$ and that the function ξ_{t-1} is well defined. Then define

$$\gamma(\xi_{t-1}(i)) = \gamma(\xi_{t-1}(i-1)) - \varepsilon$$

and

$$\gamma(\xi_{t-1}(i)) = \gamma(\xi_{t-1}(i-1)) - \varepsilon \quad \text{for } i = 2, ..., n_{t-1}.$$

Also define

$$W(\xi_{t-1}(1)) = W(\xi_t(n_t)) + \gamma(\xi_t(n_t))(\eta(\xi_{t-1}(1)) - \eta(\xi_t(n^t))) - \delta$$

and

$$W(\xi_{t-1}(i)) = W(\xi_{t-1}(i-1)) + \gamma(\xi_{t-1}(i-1)) \\ \times (\eta(\xi_{t-1}(i)) - \eta(\xi_{t-1}(i-1))) - \delta \quad \text{for } i = 2, ..., n_{t-1}.$$

We can repeat the construction up to $W(\xi_0)$, $\gamma(\xi_0)$. Since there are finitely many nodes sufficiently small δ, ε can be found to ensure that for all t and all $\xi, \zeta \in \mathcal{N}_t$, $\eta_\xi \neq \eta_\zeta$. Furthermore, since $W(\cdot)$ is constructed as a piecewise linear increasing and concave function inequalities (1) must hold.   □

With this lemma, the proof of the theorem is very short.

Proof of Theorem 1. For any $\varepsilon > 0$ and $\bar{\gamma} > 0$, if for all $\xi \in \Omega$, $\gamma(\xi) \geq \bar{\gamma}$, one can find $(\lambda(\xi))_{\xi \in \Omega}$ which solve (1) and which satisfy $0 < \lambda(\xi) < \varepsilon$. This follows from the absence of arbitrage and the fact that we can choose $\lambda(\xi_0)$ without any restrictions. Therefore, for any $\delta > 0$ and any observation on spot prices and consumptions one can find $\lambda(\xi)$ which satisfy (1) and for which

$$\sup_{\xi, \zeta \in \Omega} |\lambda(\xi) p(\xi)(c(\xi) - c(\zeta))| < \delta.$$

But now, Lemma 2 implies that inequalities (1) must hold as well since inequalities (1) hold.   □

3.1 Interpretation of the Main Theorem

We want to argue that Theorem 1 implies that the assumption of Kreps–Porteus utility imposes no restriction on individual choice behavior.

The point is easiest to illustrate in a model with no uncertainty. In this case, we can assume that one observes the behavior of an individual throughout his lifetime and that there is a unique terminal node. There is a unique $\zeta \in \mathcal{J}(\xi)$ for all non-terminal $\xi \in \Xi$, $\pi(\xi) = 1$ and $\mu(\xi) = v(\zeta)$. The assumption that $\mu(\xi)$ is observable is justified if we assume that this terminal node denotes the last period of the individual's planning horizon. In this case, we know that $\mu(\xi) = 0$ and Theorem 1 immediately implies that the assumption of Kreps–Porteus utility imposes no restrictions on individual choices in markets.

Time Consistency

Following Strotz [Str56], there have been various attempts to formalize 'dynamic inconsistency of preferences', the human tendency to prefer immediate rewards to later rewards in a way that our 'long-run selves' do not appreciate (see e.g., Gul and Pesendorfer [GP01] and the references therein).

Many papers studying time-inconsistent preferences have also searched for empirical proof that people have such preferences. It follows from Theorem 1 that it is impossible to find such empirical proof from observing individuals' choices in markets.[1] Since Kreps–Porteus utility is time consistent by

---

[1] The existence of external commitment devices and experimental evidence might offer a different perspective.

construction, this immediately implies that the assumption of time consistency imposes no restriction on choices in markets. For any present-biased preference specification and any resulting observation of choices there exists a Kreps–Porteus utility function which yields exactly the same choices.

Uncertainty

In a model with uncertainty, Lemma 1 imposes a non-trivial restriction on an extended observation. Theorem 1 is not applicable to all situation since it requires that at different terminal nodes the continuation utilities are different. If the last period of the model is interpreted as the end of an agent's planning horizon, it makes sense to assume that $\mu(\xi) = \mu(\zeta) = 0$ for all terminal nodes ξ and ζ. An example now shows that under this assumption portfolio choices are restricted by the assumption of Kreps–Porteus utility.

Example 1. Consider a two-period model with two possible states in the second period. The states are numbered 0 (today), 1, 2 and the probabilities are $\pi_1 = \pi_2 = 1/2$. Assume for simplicity that there is only one good and that the price of this good is one at each node. Assume that there are two arrow securities, one paying one unit in state 1, the other paying one unit in state 2 and that $q_1 > q_2$. Suppose that $c_1 > c_2$ and that the portfolio choice satisfies $\theta_1 > \theta_2$.

The observed portfolio choice is inconsistent with Lemma 1. Since $c_1 > c_2$, by (1), $\lambda_2 > \lambda_1$. However, by (1) this implies that $q_1 < q_2$ — a contradiction.

While the example only shows that there are restrictions on portfolio choices at time $T - 1$, there might also exist restrictions at other nodes. Consider for example an economy with identical consumptions at all last period nodes. This implies that μ_ξ has to be identical for all $\xi \in \mathcal{N}_{T-1}$, i.e., in the second to last period and Example 1 can be extended to this case.

However, in general, observed consumption will be different at all terminal nodes, leading to different continuation utilities at different nodes at $T - 1$. Theorem 1 then implies that Kreps–Porteus utility only imposes restrictions on consumptions at T and portfolio choices at $T-1$ but on no other variables.

Moreover, it is clear that when period T is not the last period in the individual's planning horizon and it is impossible to observe $\mu(\xi)$ for $\xi \in \mathcal{N}_T$ there are no restrictions whatsoever on behavior.

Apart from the special case where last period's choices are restricted, the assumption of Kreps–Porteus utility therefore imposes no restrictions on intertemporal choice under uncertainty.

Observability

If one observes a household's choices throughout time it is unlikely that the weak restrictions on last period choices are actually observable. While under certainty it is conceivable that choices and prices are observable at every period, under uncertainty, one can only observe one sample path of an underlying stochastic process. One has to make stationarity assumptions on the underlying stochastic processes for prices and incomes to imbue the model with empirical content. Under a stationarity assumption, one can estimate the processes and one therefore knows prices, dividends and incomes at all nodes of the event tree. However, while prices, dividends and incomes might be stationary, the life cycle aspect of the agent's finite horizon maximization problem implies that choices are in general not stationary. Although given a finite data set, it is always possible to construct an event tree and a stationary process for prices, dividends and endowments such that the observed variables form a sample path and the assumption of stationarity of the exogenous variables cannot be refuted, it is implausible that all variables jointly follow a first-order Markov chain. Kreps–Porteus utility only imposes restrictions on last period choices under these additional stationarity assumptions.

We also assume throughout that the agent evaluates uncertain income streams according to the true (known) probabilities. While this might seem like a very strong assumption, it is standard in the applied literature and it is clear that without any assumption an agent's beliefs, Theorem 1 will become trivial. In this case the agent could put zero probability on all but one sample path. If this happens to be the observed sample path, the model is the same as under certainty.

3.2 Assumptions on the Aggregator

In order to obtain restrictions one has to make additional assumptions on the aggregator function $W(\cdot, \cdot)$. One possibility is to require that the agents' indifference curves over current consumption are identical at all nodes. For this, we assume that $W(x, z)$ can be written as $F(w(x), z)$, where $F : \mathbb{R}_+ \times \mathbb{R}_+ \to \mathbb{R}$ is assumed to be increasing and concave and where $w : \mathbb{R}_+^L \to \mathbb{R}_+$ is the concave and increasing utility function for spot consumption. We call this aggregator function weakly separable. The assumption of weak separability ensures that marginal rates of substitution between different spot commodities are not affected by different future utilities. If there is only one good, i.e., $L = 1$ this assumption does not guarantee refutability. The assumption imbues the model with empirical content for $L > 1$ because it restricts possible choices on spot markets. There are many utility functions satisfying this assumption— for example, any nesting of concave CES-utility functions will give rise to a weakly separable aggregator.

The model is now testable. In fact, choices on spot markets together with prices for commodities $(p(\xi), c(\xi))_{\xi \in \Xi}$ must satisfy the strong version of the strong axiom of revealed preferences.

Definition 3 (Chiappori and Rochet [CR87]). $(p(\xi), c(\xi))_{\xi \in \Omega}$ satisfies SSARP if for all sequences $\{i_1, ..., i_n\} \subset \Omega$

$$p_{i_1} c_{i_1} \geq p_{i_1} c_{i_2}, p_{i_2} c_{i_2} \geq p_{i_2} c_{i_3}, ..., p_{i_{n-1}} c_{i_{n-1}} \geq p_{i_{n-1}} c_{i_n}$$

implies

$$c_{i_n} = c_{i_1}, \quad \text{or} \quad p_{i_n}(c_{i_1} - c_{i_n}) > 0$$

and if for all $\xi, \zeta \in \Omega\, p(\xi) \neq p(\zeta)$ implies $c(\xi) \neq c(\zeta)$.

Chiappori and Rochet [CR87] show that in the context of static choice SSARP is necessary and sufficient for the data to be rationalizable by a smooth, strictly concave and strictly increasing utility function. In the intertemporal context, SSARP implies that choices are rationalizable by a separable (time invariant) expected utility function if asset prices or portfolio choices are unobservable.

We say that a utility function $u(\cdot)$ is time separable if it is Kreps–Porteus and if there exists a $\beta \in [0,1]$ such that the aggregator can be written as

$$W(x, y) = w(x) + \beta y.$$

The following theorem is the main result of this section.

Theorem 2. *The following statements are equivalent.*

(a) *An extended observation*

$$\mathcal{O} = ((c(\xi), \theta(\xi), d(\xi), q(\xi), p(\xi), \pi(\xi))_{\xi \in \Omega}, (\mu(\xi))_{\xi \in \mathcal{N}_T})$$

which satisfies $\mu(\xi) \neq \mu(\zeta)$ for all nodes $\xi \neq \zeta \in \mathcal{N}_T$ and $c(\xi) \in \mathbb{R}_{++}^L$ for all $\xi \in \Omega$ is rationalizable by a Kreps–Porteus utility function with weakly separable aggregator.

(b) *There are $V(\xi), U(\xi) \in \mathbb{R}_+$, $\lambda(\xi) \in \mathbb{R}_{++}$ and $\gamma(\xi) \in \mathbb{R}_{++}^2$ for all $\xi \in \Omega$ such that,*

(1) *For all $\xi \in \Omega$, $\xi \notin \mathcal{N}_T$,*

$$q(\xi)\lambda(\xi) = \gamma_2(\xi) \sum_{\zeta \in \mathcal{J}(\xi)} d(\zeta)\pi(\zeta|\xi)\lambda(\zeta) \tag{2}$$

$$\mu(\xi) = \sum_{\zeta \in \mathcal{J}(\xi)} \pi(\zeta|\xi)U(\zeta).$$

(2) *For all $\xi \neq \zeta \in \Omega$,*

$$U(\xi) \leq U(\zeta) + \begin{pmatrix} \gamma_1 \\ \gamma_2 \end{pmatrix} \left[\begin{pmatrix} V(\xi) \\ \mu(\xi) \end{pmatrix} - \begin{pmatrix} V(\zeta) \\ \mu(\zeta) \end{pmatrix} \right] \tag{3}$$

as well as

$$V(\xi) \leq V(\zeta) + \frac{\lambda(\zeta)}{\gamma_1(\zeta)} p(\zeta) \cdot (c(\xi) - c(\zeta)). \tag{4}$$

The inequality holds strict whenever $c(\xi) \neq c(\zeta)$.

(c) *The prices and spot market choices $(p(\xi), c(\xi))_{\xi \in \Omega}$ satisfy SSARP.*

(d) *There exist asset prices* $(\bar{q}(\xi))_{\xi\in\Omega}$, *incomes* $(\bar{I}(\xi))_{\xi\in\Omega}$ *and portfolio holdings* $(\bar{\theta}(\xi))_{\xi\in\Omega}$ *such that the observation* $(c(\xi), \bar{\theta}(\xi), d(\xi), \bar{I}(\xi), \bar{q}(\xi), p(\xi),$ $\pi(\xi))_{\xi\in\Omega}$ *can be rationalized by a time-separable utility function.*

Proof. The proof of Lemma 1 above implies that (a) is equivalent to (b). The additional requirement of weak separability gives rise to the set of inequalities (3) and (4). The crucial part of the proof is to show that (b) is equivalent to (c): According to Afriat's Theorem (see Chiappori and Rochet [CR87]) SSARP is necessary and sufficient for the existence of numbers $(V(\xi)), \alpha(\xi)_{\xi\in\Omega}, \alpha(\xi) > 0$ which satisfy

$$V(\xi) - V(\zeta) \leq \alpha(\zeta)p(\zeta)(c(\xi) - c(\zeta)) \tag{5}$$

for all $\xi \neq \zeta \in \Omega$, with the inequality holding strict for $c(\xi) \neq c(\zeta)$.

To show that (b) implies (c), we define $\alpha(\xi) = \lambda(\xi)/\gamma_1(\xi)$ inequality (4) then implies inequality (5).

For sufficiency, assume that there exist numbers $(V(\xi)), \alpha(\xi)_{\xi\in\Omega}, \alpha(\xi) > 0$ which satisfy (5).

We can then choose $(\gamma_1(\xi))_{\xi\in\Omega}$ small enough to ensure that inequality (3) has a solution—this follows from the same argument as in the proof of Theorem 1: We take the $V(\xi)$ as given and construct $\gamma_2(\xi)$ analogous to the number $\gamma(\xi)$ in the previous proof. Since we do not impose restrictions on $\gamma_1(\xi)$ except bounding them from above, it is easy to ensure that $(\lambda(\xi))/(\gamma_1(\xi)) = \alpha(\xi)$ by choosing $\lambda(\xi)$ sufficiently small. Equality (2) can be satisfied because all these inequalities are homogeneous in $(\lambda(\xi))_{\xi\in\Omega}$ and impose no lower bound on $\inf_{\xi\in\Omega} \lambda(\xi)$.

Finally, we have to show that (c) is equivalent to (d): The Afriatinequalities for time separable utility are particularly easy. Inequalities (4) must hold with $\gamma_1(\xi) = 1$. Equation (2) must hold with $\gamma_2(\xi) = \beta$. Therefore, the observation can be rationalized by a time-separable utility function if and only if in addition to inequality (5) we also have

$$\alpha(\xi)q(\xi) = \beta \sum_{\zeta\in\mathcal{J}(\xi)} \alpha(\zeta)\pi(\zeta|\xi)d(\zeta).$$

Since we are free to choose the $(q(\xi))$, this can always be satisfied as long as there is no arbitrage. Portfolio choices $\theta(\xi)$ and incomes $I(\xi)$ must then be chosen to ensure that the budget constraints are satisfied.   □

It is important to point out that weakly separable Kreps–Porteus utility is not observationally equivalent with time-separable utility if portfolio choices are observable. In this case, time separability puts restrictions on portfolio holdings—weakly separable Kreps–Porteus utility does not.

4 Conclusion

Assuming the existence of utility functions to explain the behavior of consumers is standard in economics. In order to imbue models which use utility functions with empirical content one would hope that by watching the behavior of individuals throughout their life, one can test the hypothesis that these individuals maximize utility. However, we show in this paper that this is only possible under additional assumptions on the utility function. Kreps–Porteus utility with a weakly separable aggregator is one class of utility functions which imposes restrictions on individual behavior. These restrictions can be formulated in a tractable way one can test a large data set for consistency with utility maximization (see Varian [Var82] for such tests). Without this additional assumption there are no restrictions and the theory cannot be tested by observing the choices of a single individual. In this case, one needs to use panel data and assume that similar individuals have identical preferences.

The situation is more complicated when there is no data on individual choices and when one has examine restrictions on aggregate data. Brown and Matzkin [BM96] show that there exist observable restrictions for the case where one can observe how aggregate consumption varies as prices and the income distribution vary. The criticism in this paper against traditional tests of utility maximization which use individual data applies to the analysis in Brown and Matzkin (which uses aggregate data) as well. In Kubler [Kub03], we extend their analysis to a multi-period model where the observations consist of a time series on aggregate data.

Acknowledgments

I would like to thank Don Brown for very helpful discussions on earlier versions of this paper as well as his constant encouragement. I also thank Pierre-André Chiappon, Yakar Kannai, Ben Polak, Heracles Polemarchakis and an anonymous referee for very valuable comments.

Kubler, F.: Is intertemporal choice theory testable? Journal of Mathematical Economics **40**, 177–189 (2004) Reprinted by permission of Elsevier.

Observable Restrictions of General Equilibrium Models with Financial Markets

Felix Kubler

University of Pennsylvania, Philadelphia, PA 19104-6297 `kubler@sas.upenn.edu`

Summary. This paper examines whether general equilibrium models of exchange economies with incomplete financial markets impose restrictions on prices of commodities and assets given the stochastic processes of dividends and aggregate endowments. We show that the assumption of time-separable expected utility implies restriction on the cross-section of asset prices as well as on spot commodity prices. However, a relaxation of the assumption of time separability will generally destroy these restriction.

Key words: General equilibrium, Incomplete financial markets, Non-parametric restrictions

1 Introduction

General equilibrium theory as an intellectual underpinning for various fields in economics is often criticized for its lack of empirical content (see for example [HH96]). While Brown and Matzkin [BM96] challenge this view by showing that there are restrictions on the equilibrium correspondence, i.e., the map from individual endowments to equilibrium prices, it is now well understood that these restrictions only arise because *individual incomes* are observable and that general equilibrium theory imposes few restrictions on aggregate quantities and prices alone. There are no restrictions on the equilibrium set [Mas77] and there are no restrictions on the equilibrium correspondence when individual incomes are not observable and when the number of agents is sufficiently large (see, e.g., [CEKP04] for a local analysis, but see also positive results in [Sny04]).

However, in models with time and uncertainty there are various natural assumptions on individual preferences which ensure that equilibrium prices cannot be arbitrary for given aggregate endowments. When agents consume after uncertainty about endowments is resolved, it is often assumed that they have von Neumann–Morgenstern utility with common beliefs. It is well known since

Borch [Bor62] that for this case, Pareto-optimality implies that agents' optimal consumptions are a non-decreasing function of aggregate resources only. With complete financial markets competitive equilibria are Pareto-optimal and possible equilibrium prices for state contingent consumption must therefore be anti co-monotone to aggregate endowments (see e.g., [LM94] or [Dan00] for a detailed analysis of this case). There exist restrictions on the equilibrium set for fixed aggregate endowments under the assumption that markets are complete and agents maximize expected utility with homogeneous beliefs.

In this paper, we take this observation as a starting point and examine how it generalizes to economies with multiple commodities per state, incomplete financial markets, heterogeneous beliefs and several time periods.

Independently of complete financial markets and Pareto-optimality of competitive equilibrium allocations the assumption of expected utility with homogeneous beliefs turns out to impose joint restrictions on aggregate variables and cross-sectional asset prices.

In a multi-period model, when households maximize time-separable expected utility, restrictions from a two period model translate immediately to restrictions at each node of the event tree. When agents' expectations are unknown and heterogeneous there are restrictions on asset prices, dividends and aggregate endowments as long as beliefs are restricted to lie in some strict subset of possible beliefs, i.e., subjective probabilities are bounded away from zero.

In the light of the theoretical literature on equilibrium restrictions these results are not necessarily surprising because time-separable expected utility is a very strong assumption. However, under this strong assumptions, our results show that there are restrictions on the equilibrium set—individual endowments are not observed (as for example in [BM96]) but equilibrium prices are restricted for given aggregate endowments. It is natural to ask whether this is a property specific to time-separable expected utility. We show that a slight relaxation of time separability is likely to destroy all these restrictions. In particular, a model where agents maximize recursive utility imposes almost no restrictions on aggregate data, even under fairly strong additional assumptions on the aggregator which ensure that individual choices in spot markets have to satisfy the strong axiom of revealed preferences.

The paper is organized as follows. In Section 2 we give a short introduction into the model and we define formally what we mean by 'restrictions'. In Section 3 we examine restrictions on the joint process of asset prices, spot prices and aggregate endowments under the assumption of time-separable expected utility. In Section 4 we argue that time-separability of the utility functions is crucial for these results.

2 The Model

We consider a standard multi-period general equilibrium model of an exchange economy with incomplete financial markets (GEI model). This is a model with several goods, uncertainty and $T + 1$ periods $t = 0, ..., T$. (For a thorough description of the model see e.g., [DS86].) We model the uncertainty as an event tree Ξ with X nodes $\xi \in \Xi$. We denote a node's unique predecessor by ξ_- and the set of its successors by $\mathcal{J}(\xi)$. This set is empty for each terminal node. We will denote the root node (the unique node without a predecessor) by ξ_0.

There are complete spot markets for L commodities at every node and we denote commodity ℓ's price at node ξ by $p_\ell(\xi)$. We assume that the price of commodity one in non-zero at each node and normalize $p_1(\xi) = 1$ for all ξ, so that the first good is the numéraire commodity.

There are J real assets which we collect in a set $\mathcal{J}$, with asset j paying $d_j(\xi)$ units of good 1 at node ξ, its price being denoted by $q_j(\xi)$. We assume that all assets are traded at all nodes (i.e., all assets are long-lived). We will remark below on how the results change when we introduce one-period securities.

There are H agents which we collect in $\mathcal{H}$. Each agent h has an endowment $e^h \in \mathbb{R}^{LX}_{++}$, his consumption set is $\mathbb{R}^{LX}_+$ and his utility function is denoted by $u^h : \mathbb{R}^{LX}_+ \to \mathbb{R}$. Agent h's portfolio holding at node ξ is denoted by $\theta^h(\xi) \in \mathbb{R}^J$ and his consumption by $c^h(\xi) \in \mathbb{R}^L_+$. In order to simplify notation we will sometimes use $\theta^h(\xi_{0-})$ to denote agent h's portfolio holding in the beginning of period 0.

We denote aggregate endowments at node ξ by $e(\xi) = \sum_{h \in \mathcal{H}} e^h(\xi)$.

We can define an equilibrium as follows.

Definition 1. A competitive equilibrium consists of good and asset prices $(p(\xi), q(\xi))_{\xi \in \Xi}$ and of an allocation $(c^h(\xi), \theta^h(\xi))_{\xi \in \Xi}^{h \in \mathcal{H}}$ such that: all markets clear and agents maximize utility, i.e., for all $\xi \in \Xi$,

$$\sum_{h=1}^{H}(c^h(\xi) - e^h(\xi)) = 0) \quad \text{and} \quad \sum_{h=1}^{H}\theta^h(\xi) = 0.$$

For all $h \in \mathcal{H}$,

$$(c^h, \theta^h) \in \underset{c \in \mathbb{R}^{LX}_+, \theta \in \mathbb{R}^{JX}}{\arg\max} \ u^h(c) \text{ s.t.}$$

$$p(\xi)c(\xi) + q(\xi)\theta(\xi) \leq p(\xi)e^h(\xi) + \theta(\xi_-)(q(\xi) + d(\xi))$$
$$\text{for all } \xi \in \Xi \theta(\xi_{0-}) = 0.$$

A necessary condition for prices to be equilibrium prices is the absence of arbitrage opportunities.

Definition 2. Prices and dividends $(p(\xi), q(\xi), d(\xi))_{\xi \in \Xi}$ preclude arbitrage if there is no trading strategy $(\theta(\xi))_{\xi \in \Xi}$ with $\theta(\xi_{0-}) = 0$ such that if we define

$$D(\xi) = \theta(\xi_-)(d(\xi) + q(\xi)) - \theta(\xi)q(\xi),$$

$D(\xi) \geq 0$ for all $\xi \in \Xi$ and $D \neq 0$.

We want to investigate in this paper which assumptions on preferences restrict competitive equilibrium prices and aggregate endowments beyond the absence of arbitrage. We will assume throughout that prices preclude arbitrage.

2.1 Observable Restrictions

Given an event tree Ξ and H agents, a process of prices, aggregate endowments and dividends $(p(\xi), q(\xi), e(\xi), d(\xi))_{\xi \in \Xi}$ is said to be *rationalizable* if there exist specifications of agents utility functions, individual endowments and individual consumptions such that these add up to aggregate endowments and such that these individual consumptions maximize the agents' utility functions subject to the budget constraints. We will also refer to rationalizable observations as consistent processes. Processes which are not rationalizable are called inconsistent.

The main contribution of this paper is to provide assumptions on preferences which ensure that the model imposes restrictions on prices given aggregate endowments and dividends.

This paper focuses on restrictions on prices and *aggregate quantities* although, in addition to observing aggregate endowments, there might also be cases where one can hope to observe individual incomes. From a theoretical point of view this is an interesting question because there are no restrictions on aggregate variables and prices without assumptions on preferences (when individual incomes are observable, there are both local and global restrictions on the equilibrium correspondence since individual income effects are observable—see [BM96, CEKP04]). The next step must be to examine standard assumptions on preferences in models with time and uncertainty and evaluate to what extent they do impose restrictions.

From a more practical perspective, stochastic processes for individual incomes are difficult to estimate. For example, the question whether shocks to income are transitory or permanent seems difficult to resolve empirically—however, the results in [CD96] indicate that quantitative predictions of asset pricing models with heterogeneous agents and incomplete markets depend crucially on the exact specification of the individual income processes. It is then important to clarify that even without any assumptions on incomes, the standard model does restrict equilibrium prices and to investigate these restrictions.

A model imposes restrictions on the entire process if there exist a process $(p(\xi), q(\xi), e(\xi), d(\xi))_{\xi \in \Xi}$, with $e(\xi) > 0$ for all $\xi \in \Xi$ which precludes arbitrage but which is not rationalizable. From a theoretical point of view it is interesting to examine restrictions on the entire process of prices, since this means examining restrictions on the equilibrium set of the economy in the tradition of Mas-Colell [Mas77].

However, when investigating observable restrictions of general equilibrium models it must be taken into account that observed time series only consist of a single sample path of dividends, aggregate endowments and prices. It is clear that without any assumptions about off sample path realizations of dividends and endowments, the model does not restrict possible observations at all. Maheswaran and Sims [MSi93] show that even the absence of arbitrage is a restriction on the entire process and does not impose restrictions on a single sample path of asset prices, dividends and aggregate endowments. Similarly, equilibrium will not impose any restrictions on a path when off-sample path variables can be picked freely.

It is standard in modern macroeconomics (see e.g., [Luc78]) to assume that aggregate endowments as well as dividends are stationary. By estimating the stochastic process for endowments and dividends one can then specify values for these variables off the sample path. In addition to examining restrictions on the entire process of asset prices we will therefore also examine if the model restricts asset prices along a sample path, given specifications for aggregate endowments and dividends at all nodes. In particular, we will give conditions under which there exists restrictions on asset prices at $t = 0$ (the root node) given only the stochastic processes for dividends and aggregate consumption.

Given the excellent quality of data on security prices, there is no reason why stochastic processes for dividends and endowments should be easier to specify than processes for asset prices. However, if individual endowments and dividends follow a Markov chain and if markets are incomplete, the *joint* process of prices, endowments and dividends will generally not be Markov. Since the theory takes prices as endogenous, it seems natural in our framework to specify as the exogenous variables dividends and aggregate consumption and to ask if the model imposes restrictions on prices.

2.2 The Role of Preferences

Under the assumption that all utility functions are strictly increasing and strictly concave any joint process of prices, aggregate endowments and dividends can be rationalized as long as prices preclude arbitrage. The model is not refutable. While this result is well known it is worth to reformulate it in our framework. In particular, it is important to point out that the result also holds true when individual consumptions are observable. The fact that the model has no empirical content in this case has therefore nothing to do with aggregation issues—it is caused by the fact that we are considering a single observation—incomes or prices do not vary exogenously.

Theorem 1. *Spot prices* $(p(\xi))_{\xi \in \Xi}$, *asset prices* $(q(\xi))_{\xi \in \Xi}$, *asset payoffs* $(d(\xi))_{\xi \in \Xi}$ *and individual consumptions* $(c^h(\xi))_{\xi \in \Xi}^{h \in \mathcal{H}}$ *can be rationalized in a GEI model with strictly increasing and concave utility functions if and only if there is no arbitrage.*

Proof. Necessity of no arbitrage follows directly from the definition. If there exists an arbitrage opportunity, the agent's maximization problem cannot have a finite solution.

Conversely, it is well known (see e.g., [MQ96]) that the absence of arbitrage is equivalent to the existence of $\lambda(\xi) > 0$ for all $\xi \in \Xi$ such that

$$q(\xi)\lambda(\xi) = \sum_{\zeta \in \mathcal{J}(\xi)} \lambda(\zeta)(q(\zeta) + d(\zeta)). \tag{1}$$

One can construct a smooth, concave and strictly increasing utility function whose derivatives with respect to $c(\xi)$ are given by $\lambda(\xi)p(\xi)$. $\qquad\square$

Despite of this theorem one might argue that multi-period general equilibrium models with several commodities and uncertainty are testable if one makes assumptions on preferences which take into account that agents face a decision problem under time and uncertainty. One extreme is to assume that markets are complete and agents maximize time-separable expected utility with homogeneous beliefs—in this case there are restrictions on asset prices. While both the assumption of homogeneous beliefs and the assumption of complete markets can be relaxed substantially, time separability of the utility function will turn out to be crucial for the existence of restrictions.

3 Time-separable Expected Utility

We assume that for each agent h,

$$u^h((c(\xi))_{\xi \in \Xi} = W^h(\xi_0),$$

where $W^h(\xi)$, the 'utility at node $\xi \in \Xi$', is recursively defined by

$$W^h(\xi) = \begin{cases} v^h(c(\xi)) + \beta^h(\xi) \sum_{\zeta \in \mathcal{J}(\zeta)} \pi^h(\zeta)W^h(\zeta) & \text{for all non-terminal } \xi, \\ v^h(c(\xi)) & \text{for all terminal } \xi \end{cases}$$

for a strictly increasing, strictly concave and differentiable function $v^h : \mathbb{R}_+^L \to \mathbb{R}$, for varying patience factors $\beta^h(\xi) > 0$ and for conditional probabilities[1] $\pi^h(\xi)$, $\xi \in \Xi$. We assume that indifference curves do not cross the axes, i.e., for all $\bar{c} \in \mathbb{R}_{++}^L$,

$$\{c \in \mathbb{R}_{++}^L : v^h(c) \geq v^h(\bar{c})\} \text{ is closed in } \mathbb{R}_{++}^L.$$

We refer to this specification as time-separable expected utility or TSEU.

---

[1] In this context conditional probabilities are numbers $0 \leq \pi^h(\xi) \leq 1$ for all $\xi \in \Xi$ such that $\sum_{\xi \in \mathcal{J}(\zeta)} \pi^h(\xi) = 1$ for all non-terminal ζ.

We restrict each probability $\pi^h(\xi)$ to lie in some strict subset of $[0,1]$ which we denote by $I_\pi(\xi)$ and we restrict each $\beta^h(\xi)$ to lie in some bounded $I_\beta(\xi) \subset \mathbb{R}_+$. In order to relate our analysis to models with homogeneous expectations and discounting we will also consider the case where all $I_\beta(\xi)$ and all $I_\beta(\xi)$ are singleton sets. In the following we will call the collection of sets $(I_\beta(\xi), I_\pi(\xi))_{\xi \in \Xi}$ restrictions on discounting and on beliefs and we will consider observations on prices, dividends and endowments together with these restrictions.

In this section we show that under TSEU restrictions on beliefs and discounting impose joint restrictions on processes of aggregate endowments and dividends as well as on the cross section of asset prices at any non-terminal node ξ and on spot commodity prices along a sample path.

Given a joint process of prices, aggregate endowments and dividends $(p(\xi), q(\xi), e(\xi), d(\xi))_{\xi \in \Xi}$, we follow Brown and Matzkin [BM96] and use a nonparametric analysis of revealed preferences to examine restrictions on prices. We derive a system of inequalities which has a solution if and only if there are preferences and individual endowments which rationalize the data.

Lemma 1. *Given restrictions* $(I_\pi(\xi), I_\beta(\xi))_{\xi \in \Xi}$, *prices* $q(\xi) \in \mathbb{R}^J$, $p(\xi) \in \mathbb{R}^L_{++}$, *aggregate endowments* $e(\xi)$ *and dividends* $d(\xi)$, $\xi \in \Xi$, *are consistent if and only if there exist* $V^h(\xi), \lambda^h(\xi) > 0$ *and* $c^h(\xi) \in \mathbb{R}^L_{++}$ *as well as* $\pi^h(\xi) \in I_\pi(\xi)$ *and* $\beta^h(\xi) \in I_\beta(\xi)$ *for all* $\xi \in \Xi$ *and all* $h \in \mathcal{H}$ *such that*

(T1) *For all* $h \in H$ *and all non-terminal* $\xi \in \Xi$,

$$q(\xi)\lambda^h(\xi) = \beta^h(\xi) \sum_{\zeta \in \mathcal{J}(\xi)} \pi^h(\zeta)\lambda^h(\zeta)(q(\zeta) + d(\zeta)). \tag{2}$$

(T2) *For all* $\xi \in \Xi$,

$$\sum_{h \in \mathcal{H}} c^h(\xi) = e^h(\xi). \tag{3}$$

(T3) *For all* $h \in \mathcal{H}$ *and all* $\xi, \zeta \in \Xi$,

$$V^h(\xi) \leq V^h(\zeta) + \lambda^h(\zeta)p(\zeta)(c^h(\xi) - c^h(\zeta)). \tag{4}$$

The inequality holds strict whenever $c^h(\xi) \neq c^h(\zeta)$.

The proofs of the lemmas can be found in the appendix.

In the following analysis, it will be useful to define probability weighted discount factors $\delta^h(\xi)$ for each node $\xi \in \Xi$ as follows:

$$\delta^h(\xi) = \begin{cases} 1 & \text{if } \xi = \xi_0, \\ \delta^h(\xi_-)\beta^h(\xi_-)\pi^h(\xi) & \text{for all } \xi \neq \xi_0. \end{cases}$$

3.1 Restrictions on Asset Prices with a Single Commodity

In this section we examine how the assumption TSEU imposes restrictions on asset prices at the root node ξ_0, given aggregate endowments and dividends at all nodes $\xi \in \Xi$. We assume that there is single physical commodity at each state.

Applying Theorem 24.8 of Rockafellar [Roc70], since cyclical monotonicity and monotonicity are equivalent for functions of only one variable, condition (T3) turns out to be equivalent to the following.

(T3') *For all $h \in \mathcal{H}$ and all $\xi, \zeta \in \Xi$,*

$$(c^h(\xi) - c^h(\zeta))(\lambda^h(\xi) - \lambda^h(\zeta)) \leq 0. \tag{5}$$

The inequality holds strict whenever $c^h(\xi) \neq c^h(\zeta)$.

Homogeneous expectations: We first consider the case where all agents have homogeneous and known beliefs and impatience factors, i.e., we can write $\delta(\xi) = \delta^h(\xi)$ h. We show that independently of T there exist restriction on asset prices at $t = 0$ given aggregate endowments and dividends at all nodes. This, of course, implies that there exist restrictions on a sample path of prices alone.

We collect the probability weighted discounted asset payoffs in a $(X-1) \times J$ matrix A with $a_{\xi j} = \delta(\xi)d_j(\xi)$ for all $\xi \in \Xi$, $\xi \neq \xi_0$. We collect the marginal utilities in a vector $\bar{m}^h \in \mathbb{R}_{++}^{X-1}$ with $\bar{m}_\xi^h = \lambda^h(\xi)/\lambda^h(\xi_0)$.

In the following, it is not important how the nodes of the event tree are numbered as long as the ith entry in the vector $\bar{m}^h$ refers to the same node as the ith row of the matrix A. In a slight abuse of notation we will use m_i to denote the ith element of m and we will use m_ξ to denote the element of m which refers to the node $\xi \in \Xi$. Denoting the transpose of A by A^T, Condition (T1) then implies that

$$q(\xi_0) = A^T \bar{m}^h.$$

We assume that there are at least two assets in the economy, that there is at least one asset with strictly positive payoffs and that all probabilities are strictly positive. In this framework the set of arbitrage-free asset prices at $t = 0$ is given by

$$\mathcal{Q} = \{q \in \mathbb{R}^J \text{ for which there exists } m \in \mathbb{R}_{++}^{X-1} : q = A^T m\}.$$

One can use condition (T3') together with a simple linear programming argument to construct arbitrage-free prices and aggregate endowments which are not rationalizable: Fixing arbitrage free prices of $J - 1$ assets, q_j^*, $j = 1, ..., J - 1$, we can maximize the price of the Jth asset under the conditions that asset prices remain in the closure of $\mathcal{Q}$. The solution to this linear programming problem will generally lead to unique marginal utilities m. These marginal utilities will not be consistent with all possible aggregate consumptions, therefore possible prices are restricted. Formally, consider the following linear program:

$$\max_{m \geq 0} q_J = \sum_{i=1}^{X-1} m_i a_{iJ},$$

$$q_j^* = \sum_{i=1}^{X-1} m_i a_{ij}, \quad j = 1, ..., J-1.$$

Generically in A the problem is in general position and therefore has a unique solution in m (see e.g., [Dan63]). Moreover (generically in A), this solution will satisfy $m_i \neq m_j$ for all $i \neq j$. Suppose for two specific ξ, ζ we have $m_\xi < m_\zeta$. There must exist a q_J^* sufficiently close to the maximal value of the problem such that $q^* = (q_1^*, ..., q_J^*) \in \mathcal{Q}$ and such that all solutions to $q^* = A^T m$, $m \gg 0$, will still satisfy $m_\xi < m_\zeta$. For all specifications of aggregate endowments which satisfy $e(\xi) < e(\zeta)$ the equilibrium inequalities (T1), (T2) and (T3') then cannot have a solution—(T2) implies that for at least one h, $c^h(\xi) < c^h(\zeta)$. For this agent (T3') cannot be satisfied since $\lambda^h(\xi) < \lambda^h(\zeta)$.

The results can be slightly sharpened if we assume that the first asset is a consol bond i.e., if $d_1(\xi) = 1$ for all $\xi \in \Xi$. For this case the above argument becomes simpler since we can restrict the marginal utilities to lie in a compact set. In particular, we have the following theorem.

Theorem 2. *Suppose that the first asset is a consol and that there exists at least one other asset, asset 2, which has the highest payoff in one single state and the lowest payoff in another, i.e., there exists ξ_{max} such that $d_2(\xi_{max}) > d_2(\xi)$ for all $\xi \neq \xi_{max}$ and there exists ξ_{min} such that $d_2(\xi_{min}) < d_2(\xi)$ for all $\xi \neq \xi_{min}$. For any aggregate endowments which satisfy $e(\xi_{min}) \neq e(\xi_{max})$ and for each price of the consol $q_1(\xi_0)$ there then exists a $q_2(\xi_0)$ which does not allow for arbitrage but cannot be rationalized by the model.*

Proof. Consider the pricing vector $m^{max} \in \mathbb{R}_{++}^{X-1}$ defined by

$$m_{\xi_{max}}^{max} = \frac{q_1 - \varepsilon \sum_{\xi \in \Xi, \xi \neq \xi_0, \xi \neq \xi_{max}} \delta(\xi)}{\delta(\xi_{max})}, \quad m_\xi^{max} = \varepsilon \text{ for all } \xi \neq \xi_{max}.$$

Let $q_2^{max} = A^T m^{max}$ for sufficiently small $\varepsilon > 0$ the system $q = A^T m$, $m \gg 0$ only has solutions which satisfy $m_{\xi_{max}} > m_\xi$ for all $\xi \neq \xi_{max}$. Therefore q_2 can only be rationalized if $e(\xi_{max}) < e(\xi)$ for all $\xi \neq \xi_{max}$. However, also consider m^{min} defined by

$$m_{\xi_{min}}^{min} = \frac{q_1 - \varepsilon \sum_{\xi \in \Xi, \xi \neq \xi_0, \xi \neq \xi_{min}} \delta(\xi)}{\delta(\xi_{min})}, \quad m_\xi^{min} = \varepsilon \text{ for all } \xi \neq \xi_{min}.$$

This time, for sufficiently small $\varepsilon > 0$ the resulting $q_2^{min} = A^T m^{min}$ can only be rationalized if $e(\xi_{min}) < e(\xi)$ for all $\xi \neq \xi_{min}$. Clearly, both prices preclude arbitrage—however, since by assumption $e(\xi_{max}) < e(\xi_{min})$, one of prices has to be inconsistent with the aggregate endowments. $\qquad\square$

Note that the theorem does not make any assumptions about the number of investors H the number of time periods or the number of nodes X. As long as there are at least two assets with strictly positive payoffs, generically in the payoff of the second asset and in aggregate endowments, there exist arbitrage free prices which cannot be rationalized.

There is a large literature in macroeconomics which argues that given observed aggregate endowments, the average returns of stocks and government bonds cannot be explained in *commonly used* dynamic general equilibrium models. While most of this literature restricts agents' relative risk aversion and uses a parametric form for preferences, it is not well understood how a relaxation of these assumptions may resolve the puzzle and explain the low observed risk-free rate at the same time—see [Koc96].

In response to this literature Constantinides and Duffie [CD96] and Krebs [Kre04] argue that the assumption of incomplete financial markets potentially enriches the pricing implications of these models and that without restrictions on individual incomes, any (no-arbitrage) equity premium can be explained when markets are incomplete. Their results are sometimes interpreted as general results about a lack of observable restrictions in models with incomplete markets.

The reason for the negative results in [CD96] and in particular Krebs [Kre04] is the following. They assume that agents face idiosyncratic shocks which are not measurable with respect to aggregate shocks and the asset payoffs. Therefore, the condition of Theorem 2 is not satisfied in their framework—there exist states $\xi \neq \xi'$ with $d_2(\xi) = d_2(\xi') = \max_\zeta d_2(\zeta)$ and the construction in the proof breaks down.

In the above analysis restrictions on discounting are needed for restrictions on period zero asset prices. However, when asset prices are observable at all nodes of the event tree one can dispense with this restrictions. Given dividends *and prices* at direct successor nodes Theorem 2 immediately implies that generically there exist restrictions on asset prices even without assumptions on discounting.

When there are one-period assets in the economy (such as one-period bonds or derivative securities) whose state-contingent payoffs are known this observation implies that there are restrictions on the prices of these assets whenever aggregate endowments at all successor nodes are known.

Restrictions under unknown beliefs and discounting: While the assumption of homogeneous beliefs helps to achieve restrictions on asset prices it is not a necessary assumption. In particular, given any lower bound on all investors' subjective probabilities $\underline{\pi} > 0$ such that $\pi^h(\xi) \geq \underline{\pi}$ and any bounds on agents' impatience $\bar{\beta}$, $\underline{\beta}$ such that

$$0 < \underline{\beta} < \beta^h(\xi) < \bar{\beta}$$

for all $\xi \in \Xi$ and all agents $h \in \mathcal{H}$ one can construct payoffs of a risky asset, $d_2(\xi)$ to ensure that some arbitrage free prices for this asset and for a consol bond at $t = 0$ cannot be rationalized. Let $\varepsilon = (\underline{\beta}\underline{\pi})^{T-1}$. Clearly, $\delta^h(\xi) > \varepsilon > 0$ for all h and all ξ.

As before, assume that the first asset is a consol with $d_1(\xi) = 1$ for all $\xi \in \Xi$. We want to construct an asset which pays extremely high dividends at one node, while paying dividends below 1 at all other nodes. Let $\xi_{\max} = \arg\max_\xi d_2(\xi)$ and let $\xi_2 = \arg\max_{\xi \neq \xi_{\max}} d_2(\xi)$. Suppose that

$$d_2(\xi_{\max}) > 1 > d_2(\xi_2)$$

and that

$$d_2(\xi_{\max}) > \frac{\bar{\beta}^T}{\varepsilon} d_2(\xi_2).$$

If aggregate consumption is high in the state where the asset's payoff is large, $\xi_{\max}$, its price cannot be arbitrarily large. The latter of the above inequalities implies that for all agents h,

$$\delta^h(\xi_{\max})d_2(\xi_{\max}) > \delta^h(\xi_2)d_2(\xi_2).$$

If aggregate endowments satisfy $e(\xi_{\max}) > e(\xi_2)$, there must be at least one agent for whom $\lambda^h(\xi_{\max}) < \lambda^h(\xi_2)$. This imposes an upper bounds on the price of the second asset relative to the price of a consol. In fact, it must be the case that

$$q_2(\xi_0) - q_1(\xi_0) < \frac{\varepsilon}{X}(d_2(\xi_{\max}) - 1).$$

Therefore, we have the following theorem.

Theorem 3. *Suppose that there are at least two assets in the economy, i.e., $J \geq 2$ and that the first asset is a consol. For any known restrictions on beliefs and discounting $(I_\pi(\xi), I_\beta(\xi))_{\xi \in \Xi}$, if $e(\xi) \neq e(\zeta)$ for at least two nodes ξ, ζ, there exist payoffs of the second asset, $(d_2(\xi))_{\xi \in \Xi}$ and asset prices $q(\xi_0) \in \mathbb{R}^J$ such that these prices preclude arbitrage but they cannot be rationalized by the model.*

While homogeneous and known expectations might appear as an overly strong assumption on preferences it is certainly reasonable to restrict possible beliefs as in Theorem 3. However, it is clear from the preceding arguments that without such restrictions any arbitrage-free price can be rationalized by the model—as $I_\pi(\xi) \to [0, 1]$ for all ξ, the set of rationalizable prices converges to the set of no-arbitrage prices.

3.2 Joint Restrictions on Prices of Assets and Commodities

We now illustrate how restrictions on λ^h impose joint restrictions on commodity—and asset prices when $L > 1$. First suppose that markets are complete and subjective probabilities are identical and known. In this case, there exists unique solution for $\lambda^h(\xi)$, $\xi \in \Xi$ in (T1). Since this solution is identical across all $h \in \mathcal{H}$, it adding up across individuals it follows from (T3) that

$$(\lambda^h(\xi)p(\xi) - \lambda^h(\zeta)p(\zeta))(e(\xi) - e(\zeta)) \leq 0.$$

The law of demand must hold for aggregate endowments across states, given the 'Debreu' prices $\lambda(\xi)p(\xi)$. Therefore, complete asset markets, together with TSEU and homogeneous beliefs lead to strong very restrictions on spot prices, even along a sample path. When asset markets are incomplete, the restrictions on spot prices are generally much weaker since the $m^h(\xi)$ are different across agents.

However, if in addition to aggregate endowments and dividends commodity prices are known at all nodes, the above analysis still goes through: Generically in dividends, there exist asset prices at $t = 0$ and nodes ξ, ζ which ensure that all solutions to (T1) must satisfy $\lambda^h(\xi) > \lambda^h(\zeta)$. In this case, adding up, (T2) implies a joint restriction on spot prices and aggregate endowments:

$$p(\xi)e(\xi) < p(\zeta)e(\zeta).$$

For general dividends, there is no guarantee that ξ and ζ always lie on one sample path. However, one can always construct dividends, probabilities and discount factors to ensure that the model therefore restricts prices and aggregate endowments along a sample path only.

4 Relaxing Time Separability: Necessary Conditions for Restrictions

In order to show that separability is the crucial assumption for obtaining restrictions we consider a slightly more general model. Instead of assuming that current utility is simply the sum of utility from current consumption and future expected utility we now assume that it is some non-linear function of the two

$$W^h(\xi) = \begin{cases} f^h(v^h(c(\xi)), \sum_{\zeta \in \mathcal{J}(\xi)} \pi^h(\zeta)W^h(\zeta)) & \text{for all non-terminal } \xi, \\ f^h(v^h(c(\xi)), 0) & \text{for all terminal } \xi, \end{cases}$$

where $f^h : \mathbb{R} \times \mathbb{R} \to \mathbb{R}$ is assumed to be differentiable, increasing and concave. This is a special case of the recursive utility specification in [KP78]. The additional assumption of weak separability of the aggregator function ensures that marginal rates of substitution between spot commodities are not affected by different future utilities. This assumption implies that choices in spot markets must satisfy the strong axiom. Therefore there are restrictions on individual choices. However, we will show below that with sufficiently many agents this assumption generally does not impose restrictions on prices and aggregate variables even when subjective probabilities are known and identical across agents—the argument will make clear why separability of the utility function is crucial.

We refer to this specification as recursive expected utility, REU.

The equilibrium inequalities: The following lemma characterizes processes of aggregate endowments, prices and dividends which are consistent with equilibrium.

Lemma 2. *Under REU, a price process $(q(\xi), p(\xi))_{\xi \in \Xi}$ is consistent with aggregate endowments $(e(\xi))_{\xi \in \Xi}$, a dividend process $(d(\xi))_{\xi \in \Xi}$ and beliefs $(\pi^h(\xi))_{\xi \in \Xi}$ for all $h \in \mathcal{H}$ if and only if there are:*

- $c^h(\xi) \in \mathbb{R}_{++}^L$ for all $h \in \mathcal{H}$ such that $\sum_{h=1}^H c^h(\xi) = e(\xi)$ for all $\xi \in \Xi$,
- $U^h(\xi), V^h(\xi) \in \mathbb{R}$, $\gamma^h(\xi) \in \mathbb{R}_{++}^2$ and $\lambda^h(\xi) \in \mathbb{R}_{++}$, for all $\xi \in \Xi$ and all $h \in \mathcal{H}$,

such that

(R1) *For all non-terminal $\xi \in \Xi$ and all $h \in \mathcal{H}$,*

$$q(\xi)\lambda^h(\xi) = \gamma_2^h(\xi) \sum_{\zeta \in \mathcal{J}(\xi)} \pi^h(\zeta)\lambda^h(\zeta)(q(\zeta) + d(\zeta)). \tag{6}$$

(R2) *For all $h \in \mathcal{H}$ and all $\xi, \zeta \in \Xi, \xi \neq \zeta$,*

$$U^h(\xi) \leq U^h(\zeta) + \begin{pmatrix} \gamma_1^h(\zeta) \\ \gamma_2^h(\zeta) \end{pmatrix} \left[\begin{pmatrix} V^h(\xi) \\ \mu^h(\xi) \end{pmatrix} - \begin{pmatrix} V^h(\zeta) \\ \mu^h(\zeta) \end{pmatrix} \right], \tag{7}$$

where

$$\mu^h(\xi) = \begin{cases} \displaystyle\sum_{\zeta \in \mathcal{J}(\xi)} \pi^h(\zeta)U^h(\zeta) & \text{for non-terminal } \xi, \\ 0 & \text{for terminal } \xi. \end{cases}$$

In addition,

$$V^h(\xi) \leq V^h(\zeta) + \frac{\lambda^h(\zeta)}{\gamma_1^h(\zeta)} p(\zeta)(c^h(\xi) - c^h(\zeta)). \tag{8}$$

The inequality holds strict whenever $c^h(\xi) \neq c^h(\zeta)$.

No restrictions without time separability: Recursive utility imposes almost no restrictions on prices and aggregate variables, even if the aggregator is weakly separable. In the last period, since $f(v(c), 0)$ will be a strictly concave function in c, our specification of preferences is equivalent to TSEU. Therefore, there will exist restrictions on asset prices in period $T - 1$ as well as on dividends, commodity prices and aggregate endowments in period T.

However, there exist no other restrictions. This can be formally stated in various different ways. If we assume that assets' dividends in the last period are all zero, the *only* restriction is that all prices at $T-1$ are zero. Alternatively and perhaps more interestingly, there will be no restrictions on any prices at $t \leq T-1$ whenever the model has more than 2 time-periods. In the statement of the theorem we will use the former formulation, the proof of the theorem shows that the two formulations are equivalent.

Theorem 4. *If there are at least as many households as commodities, $H \geq L$, and if there are no assets with payoffs at T, i.e., $d(\xi) = 0$ for all terminal $\xi \in \Xi$, there exist no restrictions on (arbitrage-free) prices, dividends and aggregate quantities $(q(\xi), p(\xi), d(\xi), e(\xi))_{\xi \in \Xi}$ and on households' subjective probabilities $(\pi^h(\xi))_{\xi \in \Xi}^{h \in \mathcal{H}}$.*

For the proof of the theorem the following lemma from Kubler [Kub04] is needed.

Lemma 3. *For any finite event tree Ξ, probabilities $(\pi(\xi))_{\xi \in \Xi}$ and positive numbers $\eta(\xi)$ for all terminal $\xi \in \Xi$ with $\eta(\xi) \neq \eta(\zeta)$ for all $\xi \neq \zeta$ there exist numbers $(U(\xi), g(\xi))_{\xi \in \Xi}$, $1 > g(\xi) > 0$ and $U(\xi) > 0$ for all $\xi \in \Xi$ as well as a number $\delta > 0$ such that*

$$U(\zeta) - U(\xi) + g(\zeta)(\eta(\xi) - \eta(\zeta)) > 0 \quad \text{for all } \zeta, \xi \in \Xi \qquad (9)$$

with

$$\eta(\xi) = \sum_{\zeta \in \mathcal{J}(\xi)} \pi(\zeta) U(\xi) \quad \text{for all non-terminal } \xi \in \Xi.$$

The proof of the lemma requires some tedious notation but the idea is straightforward: for any concave and increasing function U with derivatives between zero and one set $g = U'$ to obtain (9).

Proof of Theorem 4. To proof the theorem, we take an arbitrary (arbitrage-free) process and construct numbers to ensure that all conditions in Lemma 2 hold.

If $H \geq L$, aggregate endowments can always be decomposed into individual consumptions that satisfy the strong axiom of revealed preferences. If we assume w.l.o.g. that $H = L$, one possible construction is to assign almost all of commodity $\ell = 1, ..., L$ to agent $h = \ell$ and only some small amount $\varepsilon_\ell > 0$ to all other agents, i.e., to set

$$c_\ell^h(\xi) - e_\ell(\xi) - (H-1)\varepsilon_\ell \text{ for } h = \ell \text{ and } c_\ell^h(\xi) = \varepsilon_\ell \text{ otherwise.}$$

If all ε_ℓ are chosen to be sufficiently small, the strong axiom must hold—$p(\xi)c^h(\xi) > p(\xi)c^h(\zeta)$ will be equivalent to $c_h^h(\xi) > c_h^h(\zeta)$. Therefore there exist $c^h(\xi)$ and there exists numbers $\alpha^h(\xi) > 0$ and $V^h(\xi) > 0$ [Afr67], such that for all $h \in \mathcal{H}$,

$$V^h(\xi) \leq V^h(\zeta) + \alpha^h(\zeta)p(\zeta)(c^h(\xi) - c^h(\zeta)) \quad \text{for all } \xi, \zeta \in \Xi. \qquad (10)$$

Furthermore, one can ensure that the inequality holds strict whenever $c^h(\xi) \neq c^h(\zeta)$.

For each agent h, given $\gamma_2^h(\xi)$ for all $\xi \in \Xi$ and given $\lambda^h(\xi_0)$ the absence of arbitrage guarantees that equation (6) have at least one solution in $\lambda^h(\xi_0)$, $\xi \in \Xi$. Given this solution define $\gamma_1^h(\xi) = \lambda^h(\xi)/\alpha^h(\xi)$ for all non-terminal

$\xi \in \Xi$. For all terminal nodes ξ, $\lambda^h(\xi)$ is arbitrary since $q(\xi) + d(\xi) = 0$. Therefore, for any terminal ξ, ζ we can choose $\gamma_1^h(\xi) > \gamma_1^h(\zeta)$ whenever $V^h(\xi) < V^h(\zeta)$ and then set $\lambda^h(\xi) = \alpha^h(\xi)\gamma_1^h(\xi)$. With these constructions construction inequalities (8) of (R2) are satisfied for all $\xi, \zeta \in \Xi$.

Since $\lambda(\xi_0)$ can be chosen freely, for each $\delta_1 > 0$, if all $\gamma_2^h(\xi)$ are bounded below, there is a solution to equation (6) with $\lambda(\xi) < \delta_1$ for all non-terminal $\xi \in \Xi$. Therefore for all $\delta_2 > 0$, we can find a $\delta_1 > 0$ to ensure that $\sup_{\xi, \zeta \in \Xi} \left\| \gamma_1^h(\xi)(V^h(\zeta) - V^h(\xi)) \right\| < \delta_2$ (for all terminal $\xi \in \Xi$ we can make all $\gamma_1^h(\xi)$ arbitrarily small without affecting any inequality). Since the $V^h(\xi)$ in equation (10) can be chosen in an open neighborhood we can ensure that $\mu(\xi) \neq \mu(\zeta)$ for all ξ, ζ and, by Lemma 3, we can construct $(U^h(\xi))_{\xi \in \Xi}$ to ensure that

$$0 < U^h(\zeta) - U^h(\xi) + \gamma_2^h(\zeta)(\mu^h(\xi) - \mu^h(\zeta)) \quad \text{for all non-terminal } \xi, \zeta.$$

We chose $\gamma_1(\xi)$ to ensure that inequality (7) holds for all terminal ξ, ζ. With the above construction the inequality now also has a solution for all non-terminal $\xi, \zeta \in \Xi$. Finally, we can ensure that the inequality has a solution for terminal and non-terminal nodes because for all terminal ξ the $\gamma_2(\xi)$ are arbitrary. □

Acknowledgments

I thank Don Brown for many helpful comments on earlier versions of this paper as well as for many valuable conversations and his constant encouragement. I thank Herakles Polemarchakis for many valuable discussions on the topic. Seminar participants at the Venice workshop in economic theory and at Yale University provided several helpful improvements. I also thank two anonymous referees who provided detailed and very helpful comments on earlier versions of the paper. Their comments improved the paper considerably. This paper contains material from my Ph.D. dissertation at Yale University. The financial support from a C.A. Anderson Prize Fellowship is gratefully acknowledged.

Kubler, F.: Observable restrictions of general equilibrium models with financial markets. Journal of Economic Theory **110**, 137–153 (2003). Reprinted by permission of Elsevier.

Appendix. Proofs of Lemmas

We first prove Lemma 2 and Lemma 1 then follows directly.

Proof of Lemma 2. For necessity, given a feasible allocation $(c^h(\xi))_{\xi \in \Xi}^{h \in \mathcal{H}}$ define

$$\mu^h(\xi) = \begin{cases} 0 & \text{for terminal } \xi, \\ \displaystyle\sum_{\zeta \in \mathcal{J}(\xi)} \pi^h(\zeta) f^h(v^h(c^h(\zeta)), \mu^h(\zeta)) & \text{for non-terminal } \xi. \end{cases}$$

Let

$$(\gamma_1^h(\xi), \gamma_2^h(\xi)) = (\partial_1 f^h(v^h(c^h(\xi)), \mu^h(\xi)), \partial_2 f^h(v^h(c^h(\xi)), \mu^h(\xi))).$$

Consider an agent's first-order condition (which are necessary and sufficient for optimality):

$$\eta^h(\xi) q_j(\xi) = \sum_{\zeta \in \mathcal{J}(\xi)} \eta^h(\zeta)(q_j(\zeta) + d_j(\zeta))$$

for $j = 1, ..., J$ and for all non-terminal $\xi \in \Xi$.

and

$$\kappa^h(\xi)\gamma_1^h(\xi)\partial_c v^h(c(\xi)) - \eta^h(\xi)p(\xi) = 0 \text{ for all } \xi,$$

where $\kappa^h(\xi_0) = 1$ and where for all $\xi \neq \xi_0 \in \Xi$,

$$\kappa^h(\xi) = \kappa^h(\xi_-)\pi^h(\xi)\gamma_2^h(\xi_-).$$

Defining $\lambda^h(\xi) = \eta^h(\xi)/\kappa^h(\xi)$ one obtains (R1). The assumption that $f^h(\cdot, \cdot)$ and $v^h(\cdot)$ are concave and the usual characterization of concave functions [Afr67] imply the inequalities in (R2)—the second optimality–condition is used to substitute for $\partial_c v^h$ in equation (8).

For the sufficiency part, assume that the unknown numbers exist and satisfy the inequalities. We can then construct a piecewise linear aggregator function f^h as well as piecewise linear $v^h(\cdot)$ following Varian [Var82]: Define

$$f^h(V, \mu) = \min_{\xi \in \Xi} \left\{ U^h(\xi) + \gamma^h(\xi) \left[\begin{pmatrix} V \\ \mu \end{pmatrix} - \begin{pmatrix} V^h(\xi) \\ \mu^h(\xi) \end{pmatrix} \right] \right\}$$

and

$$v^h(c) = \min_{\xi \in \Xi} \left\{ V^h(\xi) + \frac{\lambda^h(\xi)}{\gamma_1^h(\xi)} p(\xi)(c - c^h(\xi)) \right\}.$$

The resulting functions are concave and strictly increasing and the functions rationalize the observation. Furthermore, the approach in [CEKP04] can be used to construct a strictly concave and smooth functions v^h as well as concave and smooth f^h for all $h \in \mathcal{H}$. Their argument goes through without any modification. □

The proof of Lemma 1 now follows immediately. If for some β the aggregator can be written as $f^h(x, y) = x + \beta y$, $\gamma_1^h = 1$ and $\gamma_2^h = \beta$ will give the conditions of the lemma.

Approximate Generalizations and Computational Experiments

Felix Kubler

University of Pennsylvania, Philadelphia, PA 19104-6297 `kubler@sas.upenn.edu`

Summary. In this paper I demonstrate how one can generalize finitely many examples to statements about (infinite) classes of economic models. If there exist upper bounds on the number of connected components of one-dimensional linear subsets of the set of parameters for which a conjecture is true, one can conclude that it is correct for all parameter values in the class considered, except for a small residual set, once one has verified the conjecture for a predetermined finite set of points. I show how to apply this insight to computational experiments and spell out assumptions on the economic fundamentals that ensure that the necessary bounds on the number of connected components exist.

I argue that these methods can be fruitfully utilized in applied general equilibrium analysis. I provide general assumptions on preferences and production sets that ensure that economic conjectures define sets with a bounded number of connected components. Using the theoretical results, I give an example of how one can explore qualitative and quantitative implications of general equilibrium models using computational experiments. Finally, I show how random algorithms can be used for generalizing examples in high-dimensional problems.

Key words: Computational economics, General equilibrium, o-minimal structures

1 Introduction

Computational methods are widely used as a tool to study economic models that do not admit closed-form solutions. An important drawback of these methods is that they seem to provide information only for the specific parameter values for which the computations have been carried out. In particular, a computation can almost never prove that a model has a given property for all parameter values. The purpose of this paper is to show that, for a very wide class of economic models, to prove that a certain conclusion holds for a set of parameters that has Lebesgue measure close to 1, it is sufficient to verify the conclusion for a sufficiently large finite set of parameters. These results rest on three ideas. (1) Economic models whose descriptions involve only functions in a particular class (which is very large and contains all utility functions and production functions commonly used in applied work) give rise to sets that have very special mathematical properties. (2) In particular, deep results in algebraic geometry provide simple mechanical procedures for bounding the number of connected components of sets of parameters for which the conclusion holds. (3) From these bounds, the volume of the set can be bounded below if the set is known to contain all the points in a particular grid.

To describe these basic ideas a bit more formally, suppose the unknown set of parameters is a compact subset of Euclidean space $E \subset \mathbb{R}^\ell$. The economic conjecture is correct for an unknown set of parameters $\Phi \subset E$. Although it is not possible to use computational methods to determine that $\Phi = E$, it is often the case that for any given $\bar{e} \in E$, computational methods can determine whether $\bar{e} \in \Phi$. The question is under which conditions one can estimate the Lebesgue measure of Φ from checking that $F \subset \Phi$ for some large but finite set $F \subset E$. Obviously this is trivial if Φ is known to be convex: if a collection of points is known to lie in Φ, their convex hull must be a subset of Φ. Although it is almost never the case that Φ is convex, one can often bound the number of connected components of Φ. Koiran [Koi95] showed that from knowing upper bounds on the number of connected components of the intersection of arbitrary axes-parallel lines and the set Φ, one can construct lower bounds on the size of the set Φ by verifying that the conjecture holds on a prespecified grid $F \subset E$. The problem of proving that conjectures hold approximately thus reduces to finding bounds on the number of connected components of the set defined by the economic statement. I will argue in Section 4 that these bounds can be obtained rather mechanically from the mathematical formulation of the conjecture.

One important complication arises from the fact that numerical methods often only find approximate solutions to economic problems and that therefore it is often not possible to determine if a given $e \in E$ in fact lies in Φ or not. However, Kubler and Schmedders [KS05] argue that in many equilibrium problems, one can perform a backward error analysis and can infer from the computations that there exists a $\tilde{e}$ in a small neighborhood of e which in fact lies in Φ. In order to use this information to bound the volume of Φ, I state and prove a modified version of Koiran's result.

In order to describe the general method, a little more notation is needed. I assume that the set of unknown parameters, E is $[0,1]^\ell$ and that the economic conjecture holds true for a Lebesgue measurable set of parameters, $\Phi \subset \mathbb{R}^\ell$, which can be written in the following form

$$\Phi = \{x_0 | Q_1 x_1 Q_2 x_2, \dots, Q_n x_n ((x_0, x_1, \dots, x_n) \in X)\}, \tag{1}$$

where $Q_i \in \{\exists, \forall\}$, $x_i \in \mathbb{R}^{\ell_i}$ and X is a finite union and intersection of sets of the form

$$\{(x_0, \dots, x_n) : g(x) > 0\} \text{ or } \{(x_0, \dots, x_n) : f(x) = 0\},$$

for functions f and g in some specified class.

For a positive integer N define F to be the set of evenly spaced grid-points with distance $1/N$, i.e., $F = \{1/N, 2/N, ..., 1\}^\ell$. Suppose that for a given ε, $1/N > \varepsilon \geq 0$, the computational experiment verifies for each $e \in F$ that there is $\tilde{e}$ with $\|\tilde{e} - e\| \leq \varepsilon$ and with $\tilde{e} \in \Phi$.

Although in general one cannot rule out that there exist some $e \in E$ for which the statement is false, it might often be useful to find bounds on the size of sets of variables for which the conjecture might be wrong. Theorem 1 below shows that a bound on the number of connected components of certain subsets of Φ can be used to make a statement on the Lebesgue measure of Φ, $\text{vol}(\Phi)$. The main result of this paper is that

$$\text{vol}(\Phi) \geq 1 - \left(2\varepsilon + \frac{\ell}{N} \right) \lambda,$$

where λ is an upper bound on the number of connected components of the intersection of Φ and any axes-parallel cylinder.

Moreover, if the conjecture cannot be verified at some grid points (either because it is false at these points or because of limitations of the numerical methods used to verify the conjecture), but if the fraction of points in F at which the conjecture can be verified is some $\nu < 1$, the result implies that $\text{vol}(\Phi) \geq \nu - (2\varepsilon + \ell/N)\lambda$. In these cases, the method can still be used to estimate the volume of Φ.

The complement of Φ in $[0, 1]^\ell$ is called the exceptional set and Theorem 1 bounds its volume from above. The method will say nothing about where this set might be located and does not give lower bounds on the size of the exceptional set; in particular, it might be empty.

The resulting conclusion is, of course, much weaker than showing that the statement is true for all elements of E, but the point is that in many applications this is just not possible. The philosophy is somewhat related to the idea underlying genericity analysis for smooth economies. There one is concerned with showing that the exceptional set has measure zero. It might very well be possible that *all* of the economically relevant specifications fall into the residual set of measure zero, but it is simply not true (or cannot be shown) that the residual set is empty. Of course, there is a huge quantitative difference between showing that the residual set has small positive measure and showing that it has zero measure; in this respect generic results are much stronger than the target results in this paper.

In Section 2, I present a simple example that illustrates the basic idea of the paper. In the following two sections, I then generalize this example along two dimensions. First, the simple example assumes that the set of unknown parameters is one dimensional. In this case, it is easy to see that the number of connected components of a set tells us something about its size if one knows sufficiently many equispaced points in the set. In several dimensions this is obviously no longer true. Instead, I show in Section 3 that it suffices to work with the number of connected components of the intersection of the set and axes-parallel lines or axes-parallel cylinders.

Second, the example is constructed so that the economic conjecture can be characterized by polynomial inequalities. Although it turns out that for many economic problems the relevant functions are not always all polynomials, I will argue in Section 4 that they are very often Pfaffian (see, e.g., Khovanskii [Kho91] for definitions and motivations) and I show how the available bounds from Gabrielov and Vorobjov [GV04] on the number of connected components of the set of solutions to Pfaffian equations can be used to derive upper bounds on the number of connected components of sets that are relevant for computational experiments.

In Section 5, I discuss the applicability of these methods to general equilibrium analysis. Are there general assumptions on preferences and technologies that guarantee that all interesting statements about a given class of general equilibrium models can be tackled with the methods in this paper? Are there other classes of functions, besides Pfaffians, that guarantee the required "finiteness property"? It turns out that a necessary and sufficient conditions for these methods to be applicable to a general equilibrium model is that preferences and technology are *definable in an "o-minimal" structure*, as discussed by Blume and Zame [BZ93] or Richter and Wong [RW00]. In this context, it follows from a mathematical result on *o*-minimal structures that it is not possible to give a complete direct characterization of the class of functions for which sets defined by (1) have finitely many connected components. In this section, I also give a more elaborate example to illustrate the potential applicability of approximate generalizations to applied equilibrium analysis.

Finally, in Section 6, I discuss the computational feasibility of the method. Even with a fixed number of connected components, it turns out that the number of examples one has to compute grows exponentially with the dimension of E. Already for medium-sized problems, the methods are, therefore, often not directly applicable. An alternative is to use a random algorithm and to make statements about the size of the set of interest that are correct with high probability (see Judd [Jud97] or Blum, Cucker, Shub, and Smale [BCSS98]). Using random numbers that are generated by physical processes, one can randomly draw values for the exogenous parameters and after sufficiently many draws, for any given δ, my results then imply bounds on the probability that the true residual set is less than δ. These random algorithms are applicable even for problems for which known bounds on the number of connected components are relatively large, as long as they are orders of magnitude smaller than the errors in the computations.

2 A Simple Example

The most basic comparative statics exercise in a pure exchange economy asks what happens to equilibrium prices as individual endowments change (see, e.g., Nachbar [Nac02] for a general analysis of the problem). I consider a simple example of this exercise that is supposed to illustrate the three main ideas of the methods introduced in this paper: (1) economic models often give rise to sets that are defined by polynomial inequalities, (2) one can find bounds on the number of connected components of the set of parameters for which an economic conjecture holds, and (3) these bounds imply that the set has Lebesgue measure close to 1, once one has verified that it contains all points in a finite grid.

Suppose there are two commodities and two households with endowments e^1, e^2 and with constant elasticity of substitution (CES) utility functions

$$u^1(x) = -x_1^{-2} - 64x_2^{-2}, \quad u^2(x) = -64x_1^{-2} - x_2^{-2}.$$

Consider the conjecture that for all economies with individual endowments $e^1 = (50, e)$, $e^2 = (e, 50)$, and $e \in [0, 1]$, there exist competitive equilibria for which the equilibrium price ratio of good 2 to good 1 is (locally) decreasing in e. In these economies, there is always one competitive equilibrium for which the price ratio is equal to 1. As will become clear below, the example is constructed so that for all $e \in [0, 1]$ there are in fact three competitive equilibria, one of which exhibits a decreasing price of good 2. Suppose, however, for the sake of the example, that the only thing that is known is that for many points in $[0, 1]$, an algorithm finds one equilibrium at which the price is locally decreasing in endowments. This paper shows that it is possible to infer from this that the price must be decreasing in endowments for a large set of parameters e.

Normalizing the price of good 1 to one, equilibrium can be characterized by the requirement that aggregate excess demand for the first good is 0. Defining q to be the third root of the price of good 2 and multiplying out, one obtains that this is equivalent to

$$(8e + 25)q^3 + (-2e - 100)q^2 + (2e + 100)q - 8e - 25 = 0 \qquad (2)$$

For the price to be decreasing in e, by the implicit function theorem, it must hold that

$$3(8e + 25)q^2 - 2(2e + 100)q + 2e + 100 \neq 0,$$

$$-\frac{8q^3 - 2q^2 + 2q - 8}{3(8e + 25)q^2 - 2(2e + 100)q + 2e + 100} < 0.$$

It will turn out to be useful to write this equivalently as

$$-(8q^3 - 2q^2 + 2q - 8)(3(8e + 25)q^2 - 2(2e + 100)q + 2e + 100) < 0. \qquad (3)$$

The conjecture thus defines a set $\Phi \subset E = [0, 1]$ as follows.

$$\Phi = \{e \in [0, 1] : \exists q[(8e + 25)q^3 + (-2e - 100)q^2 + (2e + 100)q$$
$$- 8e - 25 = 0 \text{ and } - (8q^3 - 2q^2 + 2q - 8)(3(8e + 25)q^2$$
$$- 2(2e + 100)q + 2e + 100) < 0]\}.$$

This paper addresses the question of whether one can bound the Lebesgue measure of this set by computing finitely many examples, i.e., by verifying $\{0, 1/N, ..., 1\} \subset \Phi$ for some finite integer N.

Note that although it is true that for almost any $\bar{e} \in (0, 1)$, if there exists a $\bar{q}$ that satisfies (2) and (3), then it must also be true in some neighborhood of $\bar{e}$, there is no easy way to determine the size of this neighborhood. Therefore, it is not straightforward to use continuity arguments to generalize finitely many examples and to bound the size of the set Φ. In fact, it is well known in numerical analysis that zeros of high-dimensional polynomials often behave extremely sensitively with respect to small changes in the coefficients (see, e.g., Wilkinson [Wil84] for a famous example).

The main idea of this paper is as follows. Suppose that for some reason, one can obtain an upper bound, κ, on the number of connected components of Φ. Then given that in one dimension connected components must be convex, it suffices to verify that $\bar{e} \in \Phi$ for all $\bar{e} \in \{0, 1/N, ..., 1\}$ to know that the Lebesgue measure of Φ is at least $(1 - 1/N(\kappa - 1))$. The set for which the conjecture is wrong can at most be the union of $\kappa - 1$ intervals of the form $(i/N, (i+1)/N), 0 \leq i \leq 1N - 1$. Once one knows κ, one can therefore verify that the conjecture is "approximately correct" by checking it at finitely many points. Furthermore, if the conjecture can be verified only at M of the $N + 1$ points in the grid, the Lebesgue measure of Φ can still be bounded to be $M/(N + 1) - 1/N(\kappa - 1)$.

Why should it be any easier to find bounds on the number of connected components of Φ than to bound Φ by more direct arguments? The answer lies in the fact that one can bound the number of zeros of a polynomial system of equations by simply knowing the degree of the polynomials: a univariate polynomial of degree d has at most d zeros; the classical Bézout's theorem generalizes this to higher dimensions.

In Section 4, I will give rather mechanical recipes for bounding the number of connected components. For illustrative purposes, I now show in some detail how such a bound can be obtained in this example from the simple fact that a univariate polynomial of degree d has at most d zeros. It is also possible to apply the results from Section 4.

The first observation is that by the definition of Φ, equilibrium prices change monotonically in e for all $e \in \Phi$. Therefore, the number of connected components of Φ is bounded by 1 plus the number of real 0's of the two equations

$$(8e + 25)q^3 - (2e + 100)q^2 + (2e + 100)q - 8e - 25 = 0$$
$$-(8q^3 - 2q^2 + 2q - 8)(3(8e + 25)q^2 - 2(2e + 100)q + 2e + 100) = 0.$$

Moreover, by symmetry we know that for any $e \in \Phi$ there exists an equilibrium with $q = 1$ at which prices do not change. Therefore, we can factor $(q - 1)$ in both of the above equations and obtain the system

$$8eq^2 + 6eq + 8e + 25q^2 - 75q + 25 = 0$$
$$(24eq^2 - 4eq + 2e + 75q^2 - 200q + 100)(4q^2 + 3q + 4) = 0$$

For all $q > 0$, we can isolate e in the first equation and substitute it into the second to obtain the equation only in q:

$$q^4 - 2q^3 + 2q - 1 = 0. \tag{4}$$

Because this equation has at most four zeros, the number of connected components of Φ is bounded by 5. This implies that in this example, by computing equilibrium at 101 equi-spaced points and verifying that at each computed equilibrium the price is decreasing in the endowment, one can prove that the Lebesgue measure of endowments in $[0, 1]$ for which this must be true is no smaller than 0.96.

Now what happens if one can only approximate the solution to equation (2), in the sense that one finds a $\tilde{q}$ for which aggregate excess demand is approximately equal to 0, i.e., for which

$$|(8e + 25)\tilde{q}^3 + (-2e - 100)\tilde{q}^2 + (2e + 100)\tilde{q} - 8e - 25| = \varepsilon$$

for some small $\varepsilon > 0$. Although one cannot, in general claim that there exists a true equilibrium close to $\tilde{q}$, one can claim that $\tilde{q}$ is an exact equilibrium for some $\tilde{e}$ close to e. In fact,

$$\tilde{e} = \frac{25\tilde{q}^3 - 100\tilde{q}^2 + 100\tilde{q} - 25 \pm \varepsilon}{8 - 100\tilde{q} + 100\tilde{q}^2 - 8\tilde{q}^3}.$$

Given $\tilde{q}$, it is straightforward to compute bounds on $|\tilde{e} - e|$. Therefore, even if equilibrium cannot be computed exactly, one can use computational methods to verify that there are $e_0, ..., e_N$ with $e_i \in \Phi$ and $\|e_i - i/N\| < \delta$ for some small δ, $i = 0, ..., N$. This suffices to apply the method above and to bound the volume of Φ. It is easy to see that the argument goes through as before with the only modification being that now there can be four intervals of the form $(i/N - \delta, (i + 1)/N + \delta)$ that might not be subsets of Φ. Therefore, for $N = 100$, the lower bound on the volume of Φ is now $0.96 - 8\delta$.

3 Connected Components in Several Dimensions

The goal is to give good lower bounds on the size (Lebesgue measure) of Φ as defined by equation (1) in the Introduction. In this section, I consider an arbitrary (Lebesgue measurable) set $\Phi \subset \mathbb{R}^\ell$ and assume that for some reason one can obtain bounds on the number of connected components of the intersection of this set and arbitrary axes-parallel lines in $\mathbb{R}^\ell$. I will then show in Section 4 how these bounds arise from equation (1) if one limits the functions that define Φ to be of a particular class.

Throughout, fix $\| \cdot \|$ to denote the 2-norm. Define a generalized indicator function $\mathcal{J}^\varepsilon(x)$ to be 1 if there is a $y \in \Phi$ with $\|y - x\| \leq \varepsilon$ and 0 otherwise. For $\varepsilon = 0$, this is the simple indicator function and the Lebesgue measure of Φ is given by $\int_{[0,1]^\ell} \mathcal{J}^0(x)dx$.

For $x \in F$, define a cylinder of radius ε centered around $(x_1, ..., x_{\bar{i}-1}, x_{\bar{i}+1})$ by

$$C^\varepsilon_{-\bar{i}}(x) = \{y \in \mathbb{R}^\ell : \|y_i - x_i\| \leq \varepsilon \text{ for } i \neq \bar{i}\}.$$

Note that $C^0_{\bar{i}}(x)$ is simply a line parallel to the $x_{\bar{i}}$ axis passing through the point x. For a set A, denote by $\kappa(A)$, the number of its connected components.

The following lemma generalizes Lemma 2 in Koiran (1995).

Lemma 1. *Given $\bar{x} \in F$, define $Q = C^\varepsilon_{-1}(\bar{x}) \cap \Phi$. Then*

$$\left| \int_0^1 \mathcal{J}^\varepsilon(y, \bar{x}_2, ..., \bar{x}_\ell)dy - \frac{1}{N} \sum_{i=1}^N \mathcal{J}^\varepsilon\left(\frac{i}{N}, \bar{x}_2, ..., \bar{x}_\ell\right) \right| \leq \kappa(Q)/N.$$

Proof. The number of connected components of the set of $x \in C^0_{-1}$ for which $\mathcal{J}^\varepsilon(x) = 1$ is not larger than the number of connected components of $\Phi \cap C^\varepsilon_{-1}$. Therefore, it can be written as the union of K disjoint connected pieces with $K \leq \kappa$, i.e., there exist $a_1 < b_1 < \cdots < a_K < b_K$ such that

$$\left\{ x \in C^0_{-1} : \mathcal{J}^\varepsilon(x) = 1 \right\} = \bigcup_{k=1}^K \left\{ x \in C^0_{-1}, x_1 \in [a_k, b_k] \right\}.$$

Then

$$\left| \frac{1}{N} \sum_{i=1}^N \mathcal{J}^\varepsilon\left(\frac{i}{N}, \bar{x}_2, ..., \bar{x}_\ell\right) - \int_0^1 \mathcal{J}^\varepsilon(y, \bar{x}_2, ..., \bar{x}_\ell)dy \right|$$

$$\leq \sum_{k=1}^K \left| \int_{a_k}^{b_k} \mathcal{J}^\varepsilon(y, \bar{x}_2, ..., \bar{x}_\ell)dy - \frac{1}{N} \sum_{i:a_k \leq i/N \leq b_k} \mathcal{J}^\varepsilon\left(\frac{i}{N}, \bar{x}_2, ..., \bar{x}_\ell\right) \right|$$

The definition of $\mathcal{J}^\varepsilon$ implies that for all k

$$\left| \int_{a_k}^{b_k} \mathcal{J}^\varepsilon(y, \bar{x}_2, ..., \bar{x}_\ell)dy - \frac{1}{N} \sum_{i:a_k \leq i/N \leq b_K} \mathcal{J}^\varepsilon\left(\frac{i}{N}, \bar{x}_2, ..., \bar{x}_\ell\right) \right| \leq 1/N.$$

The result follows directly from this by adding up the k pieces.   □

This lemma is now extended to several dimensions. The underlying idea is to bound the number of connected components of the intersection of Φ and *any* axes-parallel cylinder. For this, define λ to be the maximal number of connected components across all intersections of Φ with all possible cylinders C^ε, i.e.,

$$\lambda = \sup_{i=1,\ldots,\ell; x \in [0,1]^\ell} \kappa(C^\varepsilon_{-i}(x) \cap \Phi). \tag{5}$$

In Section 5.2 below, I will characterize the sets for which $\lambda < \infty$. Note that it does not suffice to consider bounds on the number of connected components only across lines that connect grid points. As will become clear subsequently, the method crucially rests on the existence of a uniform bound across all possible cylinders.

The following theorem is the main tool for the analysis in this paper.

Theorem 1. *Given a bound on connected components* λ, *one can estimate the size of* Φ *by verifying that the grid* $F = \{1, \ldots, N\}^\ell \subset \Phi$ *as*

$$\left| \frac{1}{N^\ell} \sum_{i_1,\ldots,i_\ell}^N \mathcal{J}^\varepsilon \left(\frac{i_1}{N}, \ldots, \frac{i_\ell}{N} \right) - \int_{[0,1]^\ell} \mathcal{J}^0(x) dx \right| \le \left(\frac{\ell}{N} + 2\varepsilon \right) \lambda. \tag{6}$$

Proof. The theorem is proved by induction. For $\ell = 1$, one only needs to modify the last step of the proof of Lemma 1 to obtain

$$\left| \frac{1}{N} \sum_{i=1}^N \mathcal{J}^\varepsilon \left(\frac{i}{N} \right) - \int_{[0,1]} \mathcal{J}^0(x) dx \right| \le \lambda \left(\frac{1}{N} + 2\varepsilon \right).$$

For $\ell > 1$, the induction goes as follows. Adding and subtracting the term $\int_{[0,1]} (1/N^{\ell-1}) \sum_{i_2,\ldots,i_\ell}^N \mathcal{J}^\varepsilon(x_1, i_2/N, \ldots, i_\ell/N) dx_1$ to the left-hand side of equation (6), one obtains

$$\left| \frac{1}{N^\ell} \sum_{i_1,\ldots,i_\ell}^N \mathcal{J}^\varepsilon \left(\frac{i_1}{N}, \ldots, i_\ell N \right) - \int_{[0,1]^\ell} \mathcal{J}^0(x) dx \right|$$

$$\le \left| \int_{[0,1]} \frac{1}{N^{\ell-1}} \sum_{i_2,\ldots,i_\ell}^N \mathcal{J}^\varepsilon \left(x_1, \frac{i_2}{N}, \ldots, \frac{i_\ell}{N} \right) dx_1 - \int_{[0,1]^\ell} \mathcal{J}^0(x) dx \right|$$

$$+ \left| \frac{1}{N^\ell} \sum_{i_1,\ldots,i_\ell}^N \mathcal{J}^\varepsilon \left(\frac{i_1}{N}, \ldots, \frac{i_\ell}{N} \right) - \int_{[0,1]} \frac{1}{N^{\ell-1}} \sum_{i_2,\ldots,i_\ell}^N \mathcal{J}^\varepsilon \left(x_1, \frac{i_2}{N}, \ldots, \frac{i_\ell}{N} \right) dx_1 \right|.$$

Assuming that (6) holds for $\ell - 1$, one obtains that for all $x_1 \in [0,1]$,

$$\left| \frac{1}{N^{\ell-1}} \sum_{i_2,\ldots,i_\ell}^N \mathcal{J}^\varepsilon \left(x_1, \frac{i_2}{N}, \ldots, \frac{i_\ell}{N} \right) - \int_{[0,1]^{\ell-1}} \mathcal{J}^0(x_1, \tilde{x}) d\tilde{x} \right| \le \lambda \left(2\varepsilon + \frac{\ell-1}{N} \right).$$

By Lemma 1,

$$\left| \frac{1}{N^\ell} \sum_{i_1,\ldots,i_\ell}^N \mathcal{J}^\varepsilon \left(\frac{i_1}{N}, \ldots, \frac{i_\ell}{N} \right) - \int_{[0,1]} \frac{1}{N^{\ell-1}} \sum_{i_2,\ldots,i_\ell}^N \mathcal{J}^\varepsilon \left(x_1, \frac{i_2}{N}, \ldots, \frac{i_\ell}{N} \right) dx_1 \right| \le \frac{\lambda}{N}.$$

The result then follows by integrating the first term over $[0, 1]$ and adding the result to the second expression.   □

Koiran [Koi95] considered the (important) special case $\varepsilon = 0$. With bounds on the number of connected components of the intersection of Φ with axesparallel lines, this provides a method for bounding the measure of Φ. In practice, these bounds are often orders of magnitude better than bounds on connected components of the intersection with general cylinders C^{ε}. However, these bounds are applicable only in cases where the economic model can be solved exactly at the prespecified points in F; I give an example in Section 5. It is unclear, under which conditions the bounds in the theorem are tight and whether the choice of the grid points is optimal. In particular, the question of whether one can find locations of points in higher dimension that do not require the number of points to grow at the exponential rate of equation (6) is subject to further research.

Figure 1 illustrates the basic idea behind the theorem. For simplicity, consider the case $\varepsilon = 0$ and $\ell = 2$. To make the idea more transparent, it is useful to assume that the grid is in fact $\{0, 1/N, ..., 1\}^2$, i.e., includes points with one coordinate being 0.[1] Suppose $N = 3$, i.e., the conjecture can be verified on a grid of 4×4 points in $\mathbb{R}^2$ (the black dots in the figure) and suppose $\lambda = 2$. Clearly, along each horizontal line that connects grid points, there are at most two points that are not connected. The upper part of the figure depicts a generic example. No matter where the exceptional set is located, it is either the case that $2/3$ of arbitrary vertical lines cut at least three horizontal lines (as in the figure) or that $1/3$ of vertical lines cut four horizontal lines. The fact that the number of connected components along each vertical line is at most 2 now implies that there must be a set of Lebesgue measure not smaller than $2/9$ for which the conjecture holds. This is the crucial step of the argument: It seems that just from the knowledge that the conjecture holds at the grid points one can say nothing about an arbitrary vertical line that does not pass through any grid points. However, given that any horizontal line that passes through the grid points has at most one "opening" (i.e., one interval of length $1/3$ that does not lie in Φ), it follows that a large fraction of vertical lines must pass through Φ. The lower part of the figure depicts the "worst-case scenario" where the measure is, in fact, equal to $2/9$.

Although in one dimension, there is a clear relationship between convexity of a set and the set consisting of only one connected component, this is no longer true in higher dimensions. The theorem and the example show that the correct generalizations in higher dimensions consider the number of connected components along arbitrary axes-parallel lines.

In applying these method, the "only" challenge is to find reasonable bounds on λ. It turns out that computational experiments in economics usually con-

---

[1] This makes the formal proof more complicated, but helps make this specific example understandable

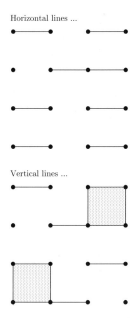

Fig. 1. An illustration of Theorem 1

sider very specific mathematical environments, for which it is easy to obtain bounds.

4 Bounding the Number of Connected Components in Economic Applications

So far, it has been assumed that bounds on the number of connected components exist and can be computed relatively easily from (1). Of course, there are many functions f, g for which the number of connected components of a set defined as in (1) might be infinite (consider, for example, the set $\{x \in (0,1) : \sin(1/x) = 0\}$) or for which it is not easily possible to compute bounds on the number of connected components.

However, in many economic application, the functions f and g in equation (1) can be written as so-called Pfaffian functions. These are classes of functions for which it can be shown that Φ has finitely many connected components. In fact, there is a fairly large literature in mathematics now that considers the problem of finding reasonable bounds on the number of connected components of sets defined by Pfaffian functions (see, e.g., Gabrielov and Vorobjov [GV04] for an overview).

4.1 Pfaffian Functions

The following definition is from Khovanskii [Kho91], who showed that these functions maintain many of the finiteness properties of polynomials.

Definition 1. A Pfaffian chain of order $r \geq 0$ and degree $\alpha \geq 1$ in an open domain $G \subset \mathbb{R}^n$ is a sequence of analytic functions $f_1, ..., f_r$ on G that satisfy differential equations

$$df_j(x) = \sum_{1 \leq i \leq n} g_{ij}(x, f_1(x), ..., f_j(x)) dx_i$$

for $\ell \leq j \leq r$. The g_{ij} are polynomial in $x = (x_1, ..., x_n)$, $y_1, ..., y_j$ of degree not exceeding α. A function $f(x) = p(x, f_1(x), ..., f_r(x))$, with p being a polynomial of degree β, is called a Pfaffian function of order r and degree (α, β).

Polynomials are included in this definition as Pfaffian functions of order 0. The following simple facts about Pfaffian functions are easy to verify.

- The expression $\exp(x)$ is a Pfaffian function of order 1 and degree $(1, 1)$ in $\mathbb{R}$; $f(x) = \log(x)$ is a Pfaffian function of order 2 and degree $(2, 1)$ on $\mathbb{R}_{++}$ because $f'(x) = 1/x$ and $f''(x) = -(f(x))^2$. Similarly, $f(x) = x^\alpha$ is a Pfaffian function of order 2 because $f'(x) = \alpha 1/x f(x)$.
- Given two Pfaffian functions of order r with the same underlying chain and degrees (α_1, β_2) and (α_2, β_2), respectively, the sum is a Pfaffian function of order r and degree $(\max(\alpha_1, \alpha_2), \max(\beta_1, \beta_2))$. The product of the two functions is Pfaffian of order r and degree $(\max(\alpha_1, \alpha_2), \beta_1 + \beta_2)$.
- A partial derivative of a Pfaffian function of order r and degree (α, β) is a Pfaffian function with the same Pfaffian chain of order r and degree $(\alpha, \alpha + \beta - 1)$.

These facts show that all commonly used utility functions and production functions (e.g., CES), as well as first order conditions for agents' optimality, can be written as Pfaffian functions.

4.2 Bounds

Gabrielov, Vorobjov, and Zell [GVZ03] showed how to compute bounds on the number of connected components of sets of the form (1). However, in their general approach it is often rather difficult to obtain good bounds because they, in fact, bound the sum of all Betti numbers Φ, while the number of connected components equals the zeroth Betti number. It is therefore useful to consider the special case of equation (1), which is often relevant in economic problems, i.e.,

$$\Phi = \{x_0 | \exists x_1 (g(x_0, x_1) > 0 \text{ and } f(x_0, x_1) = 0)\}, \tag{7}$$

where $x_0 \in \mathbb{R}^{\ell_0}, x_1 \in \mathbb{R}^{\ell_1}$, $f : \mathbb{R}^{\ell_0 + \ell_1} \to \mathbb{R}^{J_1}$ and $g : \mathbb{R}^{\ell_0 + \ell_1} \to \mathbb{R}^{J_2}$ are Pfaffian functions.

Because projection is continuous, the number of connected components of Φ is bounded by the number of connected components of

$$\{(x_0, x_1) : g(x_0, x_1) > 0 \text{ and } f(x_0, x_1) = 0\}.$$

Strict inequalities can be turned into equalities in the following way. Given J_1 inequalities $g_1(x) > 0, ..., g_{J_1}(x) > 0$ and a system of equations $f(x) = 0$ the number of connected components of

$$\{x : f(x) = 0 \text{ and } g_1(x) > 0 \text{ and } \ldots \text{ and } g_{J_1}(x) > 0\}$$

is bounded by the number of connected components of

$$\{x : f(x) = 0 \text{ and } g_1(x) \neq 0 \text{ and } \ldots \text{ and } g_{J_1}(x) \neq 0\}$$

which is bounded by the number of connected components of

$$\left\{ (x, \gamma) : f(x) = 0 \text{ and } 1 - \gamma \prod_{j=1}^{J_1} g_j(x) = 0 \right\}.$$

Given these results, it is interesting to obtain bounds on the number of connected components of sets of the form

$$S = \{x : f(x) = 0\} \subset \mathbb{R}^{\ell}, \quad f : \mathbb{R}^{\ell} \to \mathbb{R}^n$$

Suppose all f_i, $1 \leq i \leq n$ are Pfaffian functions on a domain $G \subset \mathbb{R}^{\ell}$, with either $G = \mathbb{R}^{\ell}$ or $G = \mathbb{R}^{\ell}_{++}$, having common Pfaffian chain of order r and degrees (α, β_i) respectively. Let $\beta = \max_i \beta_i$. Then the number of connected components of $\{x : f_1(x) = \cdots = f_n(x) = 0\}$ does not exceed

$$2^{\frac{r(r-1)}{2+1}} \beta(\alpha + 2\beta - 1)^{\ell-1}((2\ell - 1)(\alpha + \beta) - 2\ell + 2)^r \tag{8}$$

This bound is from Gabrielov and Vorobjov [GV04]. It grows exponentially fast in the length of the underlying Pfaffian chain and in the dimension. There is a large gap between these upper bounds and known lower bounds but for the general case. These are, to the best of my knowledge, the best currently known bounds.

Much better bounds are known for the special case where all f_i are polynomials (i.e., $r = 0$). In many economic applications, it is often sufficient to consider polynomials and it is therefore of practical importance to have good bounds for this case.

The following bound is from Rojas [Roj00]. Suppose $f_1, ..., f_n$ are polynomial and $G = \mathbb{R}^{\ell}$. Consider the convex hull of the union of the ℓ unit vectors in $\mathbb{R}^{\ell}$ together with the origin and the exponents of all monomials in the equalities which define S (i.e. for the monomial $x_1^{\alpha_1}...x_\ell^{\alpha_\ell}$ one would take the vector $\alpha \in \mathbb{R}^\ell$). For a set $Q \subset \mathbb{R}^\ell$ denote by $\mathrm{vol}(Q)$ the ℓ-dimensional volume which is standardized to obtain volume 1 for the ℓ-dimensional simplex. Then the number of connected components of S, $\kappa(S)$ can be bounded as

$$\kappa(S) \leq 2^{\ell-1}\mathrm{vol}(Q). \tag{9}$$

5 Computational Experiments in General Equilibrium Analysis

In applied general equilibrium analysis, numerical methods are routinely used to investigate quantitative features of general equilibrium models. It is therefore interesting to investigate the extent to which the methods in this paper can contribute to the current state of the art in this field.

I first discuss the extent to which computational methods can be used to verify a conjecture for a given specification of a general equilibrium model, taking into account computational errors. I then describe conditions on the fundamentals of the economy that ensure that the methods of this paper are applicable to general equilibrium models. Finally, I give an application of the methods to an example from the literature.

5.1 Approximate Competitive Equilibrium

It is useful to discuss in some detail one special case of (1). I assume that the economic statement of interest for a given specification of exogenous variables, e, can be written as

$$\exists (x_1, ..., x_k) \in \mathbb{R}^n,\ h_1(x_1, e) = 0,$$
$$\vdots \tag{10}$$
$$h_k(x_k, e) - 0,$$
$$\psi(x_1, ..., x_k, e) > 0$$

For each $i = 1, ..., k$, the (possibly multivariate) function h_i should be understood to summarize the equilibrium conditions for a given economy, i.e., they consist of necessary and sufficient first order conditions together with market clearing or consist simply of the aggregate excess demand functions. The vector x_i is supposed to contain all endogenous variables (e.g., allocations and prices) for this economy. Different h_i correspond to different specifications of the economy, for example, h_1 could summarize the equilibrium conditions for an economy without taxes, while h_2 could indicate that some taxes are introduced to the economy. The function ψ makes the comparative statics comparisons that are of interest in the particular application.

In Section 5.3, I give a concrete example from applied general equilibrium analysis. For now, I want to focus on the abstract mathematical problem. To determine whether, for a given $\bar{e} \in E$, the statement is true, one now has to compute a solution to the nonlinear system of equations.

Most existing algorithms used in practice find only one of possibly several solutions; often the algorithms find only an approximate solution. The fact that the algorithms find only one solution limits the economic statements one can consider (for example, one can generally not make statement about all solutions), but is irrelevant for the approximate generalizations suggested in this paper.

In most cases, only approximate solutions can be obtained. As an example, consider a pure exchange economy with I agents that have endowments and utility functions $\mathcal{E} = (e^i, u^i)_{i=1}^I$. For the associated aggregate excess demand function $z_{\mathcal{E}}(\cdot)$, Scarf's [Sca67] algorithm finds, for any given $\varepsilon > 0$, a $\tilde{p}$ such that $\|z_{\mathcal{E}}(\tilde{p})\| < \varepsilon$. Evidently, $\tilde{p}$ might not be a good approximation for an exact equilibrium price. However, if $z_{\mathcal{E}}(\cdot)$ is the aggregate excess demand function for a given profile of individual endowments $e^1, ..., e^I$ and if e^1 is sufficiently large, it follows from Walras' law that $z_{(e^1+z_{\mathcal{E}}(\tilde{p}), e^2, ..., e^I, (u^i))}(\tilde{p}) = 0$. Given that $\tilde{p} \cdot z_{\mathcal{E}}(\tilde{p}) = 0$, adding $z_{\mathcal{E}}(\tilde{p})$ to an agent's endowments does not change his individual demand, but only his excess demand. In other words, $\tilde{p}$ is the exact equilibrium price for a close-by economy. This observation, which has been known at least since Postlewaite and Schmeidler [PS81] (and follows from Debreu [Deb70]), has been tied to "backward error analysis" used in numerical analysis by Kubler and Schmedders [KS05]). They showed that the idea is applicable to a wide variety of general equilibrium models, including models with production, uncertainty, and, possibly, incomplete financial markets.

The fact that only approximate solutions can be obtained then means that the computational experiment can only determine an $\varepsilon > 0$ such that there exists an $\tilde{e}$ with $\|\tilde{e} - \bar{e}\| \leq \varepsilon$ and $\tilde{e} \in \Phi$.

In some cases, if the functions h_i are all polynomials, one can apply Smale's so-called alpha method to bound the difference between true equilibrium prices and allocations, and computed prices and allocations.

Smale's Alpha Method

Because it is not very well known in economics, Smale's method is summarized for completeness. The following results are from Blum, Cucker, Shub, and Smale [BCSS98, Ch. 8]).

Let $D \subset \mathbb{R}^n$ be open and let $f : D \to \mathbb{R}^n$ be analytic. For $z \in D$, define $f^{(k)}(z)$ to be the kth derivative of f at z. This is a multilinear operator that maps k-tuples of vectors in D into $\mathbb{R}^n$. Define the norm of an operator A to be

$$\|A\| = \sup_{x \neq 0} \frac{\|Ax\|}{\|x\|}.$$

Suppose that $f^{(1)}(z)$, the Jacobian of f at z, is invertible and define

$$\gamma(z) = \sup_{k \geq 2} \left\| \frac{(f^{(1)}(z))^{-1} f^{(k)}(z)}{k!} \right\|^{1/(k-1)}$$

and

$$\beta(z) = \|(f^{(1)}(z))^{-1} f(z)\|.$$

Theorem 2. *Given a $\bar{z} \in D$, suppose the ball of radius $(1 - \sqrt{2}/2)/\gamma(\bar{z})$ around $\bar{z}$ is contained in D and that*

$$\beta(\bar{z})\gamma(\bar{z}) < 0.157.$$

Then there exists a $\tilde{z} \in D$ with

$$f(\tilde{z}) = 0 \ and \ \|\bar{z} - \tilde{z}\| \leq 2\beta(\bar{z}).$$

Note that the result holds for general analytic functions. However, it is only applicable for polynomials, because, for general analytic functions, it is difficult or impossible to obtain bounds on $\sup_{k \geq 2} \|((f^{(1)}(z))^{-1} f^{(k)}(z))/k!\|^{1/(k-1)}$. In Section 5.3, I give a trivial example where the method is applicable.

5.2 Generalizable Economies

In general equilibrium analysis, the following question arises naturally: What assumptions on fundamentals guarantee that there are bounds on the number of connected components of sets defined as in equation (1)?

If the economic conjecture can be generalized from finitely many examples to a set of large volume, I say that the economic model allows for approximate generalizations. Although it is true that CES utility and production functions are Pfaffian, one would ideally hope for assumptions on preferences and technologies that are sufficient for approximate generalizations and that are a bit more general than assuming Pfaffians. Furthermore, the question arises whether there are necessary conditions on preferences and technologies that have to hold so that the techniques in this paper are applicable and the economy allows for approximate generalizations.

One possible characterization, which will turn out to be both necessary and sufficient, is that the underlying classes of economies are *definable in an o-minimal structure*. I give a brief explanation of what this means and then discuss its implications.

O-minimal Structures

The following definitions are from van den Dries [Van99]. Define a structure on $\mathbb{R}$ to be a sequence $\mathcal{S} = (\mathcal{S}_m)_{m \in \mathbb{N}}$ such that, for each $m \geq 1$, the following statements hold:

(S1) $\mathcal{S}_m$ is a Boolean algebra of subsets of $\mathbb{R}^m$.

(S2) If $A \in \mathcal{S}_m$, then $\mathbb{R} \times A$ and $A \times \mathbb{R}$ belong to $\mathcal{S}_{m+1}$.

(S3) There exists $\{(x_1, ..., x_m) \in \mathbb{R}^m : x_1 = x_m\} \in \mathcal{S}_m$.

(S4) If $A \in \mathcal{S}_{m+1}$, then $\pi(A) \in \mathcal{S}_m$, where $\pi : \mathbb{R}^{m+1} \to \mathbb{R}^m$ is the projection map on the first m coordinates.

A set $A \subset \mathbb{R}^m$ is said to be definable in $\mathcal{S}$ if it belongs to $\mathcal{S}_m$. A function $f : \mathbb{R}^m \to \mathbb{R}^n$ is said to be definable in $\mathcal{S}$ if its graph belongs to $\mathcal{S}_{m+n}$.

An o-minimal structure on $\mathbb{R}$ is a structure such that the following statements hold:

(O1) $\{(x, y) \in \mathbb{R}^2 : x < y\} \in \mathcal{S}_2$.

(O2) The sets in $\mathcal{S}_1$ are exactly the finite unions of intervals and points.

It can be easily verified that a set Φ as defined in (1) belongs to an o-minimal structure $\mathcal{S}$ if all functions f, g are definable in $\mathcal{S}$.

It is beyond the scope of this paper to discuss the assumption of o-minimality in detail. Theorem 2 below makes it clear that this assumption is very useful for the analysis. For a thorough reference on o-minimal structures, see van den Dries [Van99]. A well-known example of an o-minimal structure is induced by the ordered field of real numbers; definable sets are the semialgebraic sets. In formulation (1), the functions f and g are then all polynomials. Wilkie [Wil96] proved that the structure generated by Pfaffian functions is also o-minimal. The following two theorems are important for our analysis. The first theorem is a standard result for o-minimal structures (see, e.g., van den Dries [Van99]).

Theorem 3. *Let $\Phi \subset \mathbb{R}^\ell$ be a definable set in an o-minimal structure on $\mathbb{R}$. There is a uniform bound B such that for any affine set $L \subset \mathbb{R}^\ell$, the set $\Phi \cap L$ has at most B connected components.*

A set is affine if it can be defined by a system of linear equations.

The fact that the uniform bounds exist is interesting from a theoretical perspective, because it implies that for any set definable in an o-minimal structure, λ as defined in equation (5) is finite. In practice, however, how to obtain actual bounds when the sets cannot be described by either polynomial or Pfaffian functions is an open question.

The second result follows from the cell-decomposition theorem (see van den Dries [Van99] for a statement and proof of the cell-decomposition theorem).

Theorem 4. *If $\mathcal{S}$ is an o-minimal structure on $\mathbb{R}$, all definable sets are Lebesgue measurable.*

O-minimal Economies

Given an o-minimal structure $\mathcal{S}$, preferences $\succeq$ over consumption bundles in some definable set X are called definable if all better sets are definable, i.e., for all $x \in X$, $\{y : y \succeq x\}$ is definable in $\mathcal{S}$. Richter and Wong [RW00] proved that definable preferences can be represented by definable utility functions. It is easy to see that definable utility functions give rise to definable best response correspondences and that in pure exchange economies the equilibrium manifold is definable if preferences are definable. Blume and Zame [BZ93] applied o-minimality to consumer theory and general equilibrium analysis, and proved a definable analogue of Debreu's theorem on generic local uniqueness.

Both Blume and Zame and Richter and Wong argued that the assumption that preferences and technologies are definable in an o-minimal structure is very natural and satisfied in almost all (finite) applied general equilibrium models.

Given a class of o-minimal economies, any statement that gives rise to a definable set Φ can be approximately generalized using the methods in this paper. As mentioned above, a set Φ defined by (1) is definable in an o-minimal structure $\mathcal{S}$ if the functions f and g are definable. Moreover, any first order sentence about definable economies defines a set Φ that admits bounds on the number of connected components.

It is clear that the assumption of o-minimality of the classes of economies considered is both necessary and sufficient for the applicability of approximate generalizations. Theorem 2 shows sufficiency; necessity follows directly from the condition (O2). If an economy is not o-minimal, there exist sets with infinitely many connected components.

A Complete Characterization?

Although o-minimality is necessary and sufficient for approximate generalizations, the assumption is a bit unsatisfactory in that it only provides an indirect characterization of preferences and technologies. This leads to the question whether one can derive a largest class of utility and production functions that guarantee that the underlying economy is definable in an o-minimal structure.

Surprisingly, one can show that this is impossible. Rolin, Speissegger, and Wilkie [RSW03] constructed a pair of distinct o-minimal structures on the reals that are not both reducts of a common o-minimal expansion. This result implies that there cannot be one largest class of utility and production functions that gives rise to o-minimal economies. Instead, the assumption on ominimality is an assumption on the entire economy. If some agents' preferences are definable in one o-minimal structure while others are definable in another o-minimal structure, it is not guaranteed that there exists a larger structure that is still o-minimal and in which all preferences are definable.

In this sense, o-minimality provides the best characterization of finiteness one can hope for.

5.3 An Example

I reexamined a well-known example from applied general equilibrium analysis. Following Shoven [Sho76], I asked about the output effects of capital taxation in the two-sector model of the U.S. economy. The example is intended to give an illustration of the methods and is, therefore, held as simple as possible.

In a static economy, two consumption goods are produced by two sectors, $j = 1, 2$, using as input capital and labor. Production functions are Cobb–Douglas and of the form

$$f_1(y_{1\ell}, y_{1k}) = \gamma_1 y_{1\ell}^{2/3} y_{1k}^{1/3}$$
$$f_2(y_{2\ell}, y_{2k}) = \gamma_2 y_{2\ell}^{1/2} y_{2k}^{1/2}$$

Two individuals, $i = 1, 2$, are endowed with capital and labor (k_i, ℓ_i). Let $K = k_1 + k_2$ and $L = \ell_1 + \ell_2$ denote aggregate endowments. Utilities are Cobb–Douglas, $u^i(x_1, x_2) = x_1^{\xi_i} x_2^{1-\xi_i}$, $0 < \xi_i < 1$. Prices are (p_1, p_2, p_k, p_ℓ) and I normalize $p_\ell = 1$ throughout.

In the benchmark equilibrium, there is a tax on capital in sector 1 and the revenue $T = \tau p_k y_{1k}$ is distributed equally among the two agents. Following Shoven [Sho76], I assume that in the benchmark, equilibrium prices, as well as total output per sector are observable. The economic conjecture is that removal of this tax will increase total output, measured at new equilibrium prices, by at least 5 percent. I want to illustrate how to use the methods in this paper to prove that the conjecture is true for a large set of exogenous parameters for which the model is consistent with the benchmark equilibrium. The following five steps are necessary:

1. Formulate the economic statement as a system of equations and inequalities.
2. Identify the set of exogenous parameters that one wants to consider. This might involve adding additional constraints on parameters to match quantities in the benchmark equilibrium.
3. Formulate a system of equations that defines the set Φ and that are Pfaffian or polynomial.
4. Perform the computations on a grid of parameter values. For each parameter on the grid, either prove that it is in Φ (e.g., if Smale's method is applicable) or show that there are close-by parameters that lie in Φ.
5. Find bounds on the number of connected components and apply Theorem 1.

In step 1, note that competitive equilibrium is characterized by the following market clearing and firms' optimality conditions

$$\xi_1 \frac{p_k k_1 + \ell_1 + T/2}{p_1} + \xi_2 \frac{p_k k_2 + \ell_2 + T/2}{p_1} = \gamma_1 y_{1\ell}^{2/3} y_{1k}^{1/3}, \tag{11}$$

$$(1 - \xi_1)\frac{p_k k_1 + \ell_1 + T/2}{p_2} + (1 - \xi_2)\frac{p_k k_2 + \ell_2 + T/2}{p_2} = \gamma_2 y_{2\ell}^{1/2} y_{2k}^{1/2}, \qquad (12)$$

$$y_{1k} + y_{2k} = k_1 + k_2, \qquad (13)$$

$$y_{1\ell} + y_{2\ell} = \ell_1 + \ell_2, \qquad (14)$$

$$p_1 \frac{2\gamma_1}{3} y_{1\ell}^{-1/3} y_{1k}^{1/3} = 1, \qquad (15)$$

$$p_1 \frac{\gamma_1}{3} y_{1\ell}^{2/3} y_{1k}^{-2/3} = p_k + \tau, \qquad (16)$$

$$p_2 \frac{\gamma_2}{2} y_{2\ell}^{-1/2} y_{2k}^{1/2} = 1, \qquad (17)$$

$$p_2 \frac{\gamma_2}{2} y_{2\ell}^{1/2} y_{2k}^{-1/2} = p_k. \qquad (18)$$

In addition, the economic conjecture is that

$$\gamma_1 p_1 y_{1\ell}^{2/3} y_{1k}^{1/3} + \gamma_2 p_2 y_{2\ell}^{1/2} y_{2k}^{1/2} > 1.05 \times \text{benchmark output}. \qquad (19)$$

Note that all functions are semialgebraic and that one can rewrite these equations as a system of polynomial equations.

In step 2, the fact that output per sector is observable together with the firms' optimality conditions uniquely determine the parameters in the production functions as well as firms' factor demand. For concreteness, suppose total outputs are 60 and 40, and that benchmark prices are all equal to 1. This implies $\gamma_1 = \gamma_2 = 2$, $L = 60$ and $K = 36.875$ with a tax rate around 18.5 percent (to match the equilibrium exactly, with $\gamma_1 = \gamma_2 = 2$ one needs to set $\tau = (32,767/19,683)^{1/3}$), which corresponds to a tax revenue of $T = 3.125$. The observations impose the following restrictions on agents' preference parameters and the distribution of capital and labor endowments across agents:

$$\xi_1(k_1 + \ell_1 + T/2) + \xi_2(T/2 + K + \ell - k_1 - \ell_1) = 60. \qquad (20)$$

Without loss of generality, one can assume that agent 1 holds less capital than agent 2. Furthermore, I assume that he has larger labor endowments and a propensity to consume commodity 1 of less than $1/2$. Therefore, I define the set of admissible exogenous variables to be

$$E = [0,10] \times [30,55] \times [0.05, 0.5].$$

For each $(k_1, \ell_1, \xi_1) \in E$, the ξ_2 that solves equation (20) turns out to lie between 0 and 1.

In step 3, given the definition of E, one now needs to find a system of polynomial equations that characterizes the set Φ, i.e., the set of all $e \in E$ such that $\exists (p_1, p_2, p_k), (\bar{y}_{jk}, \bar{y}_{j\ell})_{j=1,2}$, that satisfy equations (11)–(18) with $\tau = T = 0$, $\ell_2 = L - \ell_1$, $k_2 = K - k_1$ as well as equations (19) and (20).

To find good bounds on the number of connected components, it is now useful to rewrite these equations as polynomial equations, substituting for as many variables as possible. In general, computer algebra systems such as Maple or Mathematica are ideally suited for this task. In this simple example, it can be easily done by hand. Using the fact that

$$p_k = \frac{y_{1\ell}}{2y_{1k}} = \frac{y_{2\ell}}{y_{2k}},$$

the system (11)–(18) and (19) can be rewritten as a system of polynomial equations and inequalities in $y = y_{1\ell}$ and in the unknown parameters. After some relatively straightforward algebra, one obtains that $(k_1, \ell_1, \xi_1) \in \Phi$ if there exist $y > 0$ and such that

$$-59\ell_1^2\xi_1 + 165\ell_1 k_1\xi_1 + \frac{12}{5}\ell_1 k_1 y - 192\ell_1 k_1 + \frac{17,405}{16}\ell_1\xi_1$$

$$-\frac{177}{2}\ell_1 y + 7,080\ell_1 - 96k_1^2\xi_1 + \frac{12}{5}k_1^2 y - 2,670k_1\xi_1$$

$$-\frac{1,299}{4}k_1 y + 1,1520k_1 + \frac{22125}{2}\xi_1 + \frac{278,775}{32}y - 424,800 = 0$$

$$15 - \frac{1}{2}y > 0$$

In the fourth step, one needs to verify that the conjecture is correct on a grid of points. For concreteness, I take F to consist of $1,000 \times 1,000 \times 1,000$ equispaced points,

$$F = \{0.01, 0.02, \ldots, 10\} \times \{30.025, 30.050, \ldots, 55\} \times \{0.05045, 0.0509, \ldots, 0.5\}.$$

Because y is a linear function in the parameters, it is trivial to find good error bounds. The fact that if $a + bx = \varepsilon$, there is an $\bar{x}$ with $a + b\bar{x} = 0$ and $|\bar{x} - x| = \varepsilon/b$ can be seen as a trivial special case of Smale's formula. The fraction of points that can thus be shown to lie in Φ turns out to be greater than 0.999998.

Last, in the fifth step of the procedure, to bound the size of Φ, one then needs a bound on the number of connected components. For this, one can use equation (8) from Section 4.2. How large is the number of connected components of Φ along a line parallel to the ℓ_1 axes? To bound this, one needs to fix an arbitrary $\bar{\xi}_1$, $\bar{k}_1$ and consider the set of all y, ℓ_1, t such that

$$-59\ell_1^2\bar{\xi}_1 + 165\ell_1\bar{k}_1\bar{\xi}_1 + \frac{12}{5}\ell_1\bar{k}_1 y - 192\ell_1\bar{k}_1 + \frac{17,405}{16}\ell_1\bar{\xi}_1$$

$$-\frac{177}{2}\ell_1 y + 70,80\ell_1 - 96\bar{k}_1^2\bar{\xi}_1 + \frac{12}{5}\bar{k}_1^2 y - 2,670\bar{k}_1\bar{\xi}_1$$

$$-\frac{1,299}{4}\bar{k}_1 y + 11,520\bar{k}_1 + \frac{22125}{2}\bar{\xi}_1 + \frac{278,775}{32}y - 424,800 = 0$$

$$1 - t\left(15 - \frac{1}{2}y\right) = 0$$

The volume of the convex hull of the three unit vectors together with $(2,0,0)$, $(1,1,0)$, and $(0,1,1)$ is less than 2. Therefore, one obtains a bound on the number of connected components of $2^{3-1} \times 2 = 8$.

For lines along the ξ_1 or k_1 axes, the argument is similar and one obtains $\lambda \leq 8$. To apply Theorem 1, note that $\lambda(1/N) = 24/1,000$. Therefore, the normalized volume of Φ is greater than 0.975.

6 A Random Algorithm

Suppose one has access to a random number generator and can draw uniformly and independently random $\tilde{e} \in \{1/N, ..., 1\}^{\ell}$. See, e.g., L'Ecuyer [LEc02] for a discussion on generating random and quasi-random numbers. There are now a few web sites that offer sources of random numbers that are generated by physical processes. For example, at *www.randomnumbers.info*, the user can download numbers that are generated by a physical random number generator that exploits an elementary quantum optics process. A precise description of the physical principles that underlie the method can be obtained at that site.

Both random and pseudo-random numbers are naturally integer valued (see Blum, Cucker, Shub, and Smale [BCSS98] for a more elaborate discussion on *probabilistic machines*) and, therefore, lie on a grid. It is not possible to draw random numbers uniformly over an interval, but it is possible to generate random draws from a finite set. Suppose as before that $E = [0,1]^{\ell}$ and that $F = \{1/N, ..., 1\}^{\ell}$. In this formulation, N is now the number of digits of the random numbers and can be thought of as relatively large (however, one should keep in mind that the cost of generating random numbers increases with the size of the numbers; it is not reasonable to assume that N is arbitrarily large).

Suppose one has $M \times \ell$ random numbers drawn independently and identically distributed from $\{1, ..., N\}$. Scaled appropriately, this gives M random vectors $\tilde{e}^1, ..., \tilde{e}^M \in F$. If, for each $i = 1, ...M$, there is an $\bar{e}$ with $\|\bar{e} - \tilde{e}\| \leq \varepsilon$ and with $\bar{e} \in \Phi$, $\mathcal{J}^{\varepsilon}(\tilde{e}^i) = 1$ for all i, then by the binomial formula one obtains that the probability of the event that the fraction of points $x \in F$ for which $\mathcal{J}^{\varepsilon}(x) = 0$ is greater than δ must be less than or equal to $(1-\delta)^M$. Therefore,

$$\text{Prob}\left[\frac{1}{N^{\ell}} \sum_{i_1,...,i_{\ell}} \mathcal{J}^{\varepsilon}\left(\frac{i_1}{N}, ..., \frac{i_{\ell}}{N}\right) < 1-\delta\right] \leq (1-\delta)^M.$$

Using Theorem 1, one can now infer probabilistic statements about the size of Φ from probabilistic statements about the number of points in the finite grid for which the statement is true. If $\varepsilon < 1/N$ and, as before, letting λ denote a bound on the maximal number of connected components of $\Phi \cap C_i^{\varepsilon}$, the fact that

$$\int_{[0,1]^{\ell}} \mathcal{J}^0(x)dx \geq \frac{1}{N^{\ell}} \sum_{i_1,...,i_{\ell}} \mathcal{J}^{\varepsilon}\left(\frac{i_1}{N}, ..., \frac{i_{\ell}}{N}\right) - \left(2\varepsilon + \frac{\ell}{N}\right)\lambda$$

implies that

$$\text{Prob}\left[\int_{[0,1]^\ell} \mathcal{J}^0(x)dx < 1 - \delta - \left(2\varepsilon + \frac{\ell}{N}\right)\lambda\right] \le (1 - \delta)^M. \qquad (21)$$

See Koiran [Koi95] or Blum, Cucker, Shub, and Smale [BCSS98, Ch. 17.4] for the case $\varepsilon = 0$ and a discussion of the result.

Note that the number of points needed is independent of the dimension, which enters only through bounds on the number of connected components. In the example in Section 5.3, one had to verify the conjecture at a billion points to bound the size of the residual set to be less than 0.025. Using the probabilistic approach, if one draws 1,000 random points with five significant digits (i.e., $N = 100,000$) from the set of admissible parameters and takes $\delta = 0.0049$, one obtains that the probability that the set Φ is greater than $0.996 - 3\lambda/100,000$ is at least $1 - (1 - 0.0049)^{1,000} = 0.9926$. It is, therefore, easy to verify that with high probability (at least 0.9926) the set Φ is greater than 0.995. Note that the bound on λ obtained in Section 5.3 is crucial to this argument. In the example, 1,000 points suffice to show that the set is greater than 0.995 with high probability, but one needs a billion points to show that this is true with certainty.

Although the random method is much more efficient, it is not clear how to interpret a statement like "29 is a prime number with probability 0.9926." Even though it is well known in theoretical computer science that random algorithms often reduce the complexity of the problem considerably, these algorithms usually solve a specific problem and it can often be checked that the candidate solution produced by the algorithm is an actual solution (without probabilities attached). One possible interpretation of equation (21) is the following. Suppose *nature* draws randomly a vector of parameters e uniformly from $[0,1]^\ell$. Equation (21) implies that the overall probability that this parameter will lie in Φ is at least $(1 - \delta - (2\varepsilon + \ell/N)\lambda)(1 - \delta)^N)$. This, therefore, allows for statistical statements about how likely it is that the conjecture is true for randomly selected parameters.

Note that the number of connected components can be fairly large if N is sufficiently large and ε is very small. However, in practice a bound for the volume arises naturally from the precision with which equilibria can be computed, i.e., with ε. Because Theorem 1 is only valid if $1/N > \varepsilon$, these methods are applicable if and only if the number of connected components is orders of magnitude smaller than $1/\varepsilon$.

A naive application of the random algorithm without knowledge of bounds on λ as defined in equation (5) does not allow for any statements about the size of Φ. If, for example, $\Phi \subset [0,1]$ consists of all irrational numbers, even without any computational error, because any random number will certainly be rational (in fact integer valued), the method predicts the volume of Φ to be 0, while in reality it is equal to 1. Only when one can bound λ (even if this bound turns out to be fairly large) can one make meaningful probabilistic statements about the size of Φ by randomly sampling it.

7 Conclusion

Computational experiments that make statements about one specific example economy can be generalized to infinite classes of economies when the economic fundamentals are definable in an o-minimal structure. Theorem 1, the main theoretical result, precisely specifies the conditions under which finitely many examples suffice to make statements about sets of parameters with positive Lebesgue measure.

I argue that this theoretical insight can be fruitfully put to work in applied general equilibrium analysis. For all commonly used specifications of utility and production functions, one can easily compute how many examples are needed to make statements about *large* sets of parameters. These statements are possible even if equilibria cannot be computed exactly. The methods are directly applicable to models whose solutions can be characterized by finite systems of equations; this includes stationary equilibria and steady states in infinite horizon models used in modern macroeconomics and public finance.

However, it turns out that in large problems the number of examples needed is astronomically high and it is, therefore, not feasible to make general statements using a deterministic algorithm. A random algorithm can be used to make statements about the probability that a given conjecture holds for a set of relative size $1 - \delta$.

Computing numerous random examples and then using statistical inference to summarize the findings is not a new idea (see, e.g., Judd [Jud97]), but it has not previously been formalized to take into account finite precision arithmetics of actual computations. For this method the most important practical insight of this paper is about the interplay of errors in computation, ε, the size of the random numbers used, N, and the number of connected components. One can estimate the Lebesgue measure of the set Φ by randomly drawing examples if the number of connected components is orders of magnitude smaller than $1/\varepsilon$. Otherwise, it is not possible to make even probabilistic statements about the size of Φ.

The methods introduced in this paper are obviously not the only ones that can be used to show that a given formula holds for a rich class of parameters. Because the real closed field is decidable, one can apply algorithmic quantifier elimination and use an algorithm to verify whether a given semialgebraic statement of interest is true *for all* parameters in a given (semialgebraic) set (see, e.g., Basu, Pollack, and Roy [BPR03]). However, for more complicated structures, decidability is an open problem and there are certainly no algorithms available for quantifier elimination at this time. Moreover, even for the semialgebraic case, the methods in this paper are much more tractable than quantifier elimination.

Acknowledgment

I thank seminar participants at various universities and conferences, and especially Don Brown, Dave Cass, Ken Judd, Narayana Kocherlakota, Mordecai Kurz, George Mailath, Marcel Richter, Klaus Ritzberger, Ilya Segal, co-editor, and four anonymous referees for helpful discussions and comments.

Kubler, F.: Approximate generalizations and computational experiments. Econometrica **75**, 967–992 (2007). Reprinted by permission of the Econometric Society.

Approximate Versus Exact Equilibria in Dynamic Economies

Felix Kubler[1] and Karl Schmedders[2]

[1] University of Pennsylvania, Philadelphia, PA 19104-6297 kubler@sas.upenn.edu
[2] Northwestern University, Evanston, IL 60208
  k-schmedders@kellogg.northwestern.edu

Summary. This paper develops theoretical foundations for an error analysis of approximate equilibria in dynamic stochastic general equilibrium models with heterogeneous agents and incomplete financial markets. While there are several algorithms that compute prices and allocations for which agents' first-order conditions are approximately satisfied ("approximate equilibria"), there are few results on how to interpret the errors in these candidate solutions and how to relate the computed allocations and prices to exact equilibrium allocations and prices. We give a simple example to illustrate that approximate equilibria might be very far from exact equilibria. We then interpret approximate equilibria as equilibria for close-by economies; that is, for economies with close-by individual endowments and preferences.

We present an error analysis for two models that are commonly used in applications, an overlapping generations (OLG) model with stochastic production and an asset pricing model with infinitely lived agents. We provide sufficient conditions that ensure that approximate equilibria are close to exact equilibria of close-by economies. Numerical examples illustrate the analysis.

Key words: Approximate equilibria, Backward error analysis, Perturbed economy, Dynamic stochastic general equilibrium, Computational economics

1 Introduction

The computation of equilibria in dynamic stochastic general equilibrium models with heterogeneous agents is an important tool of analysis in finance, macroeconomics, and public finance. Many economic insights can be obtained by analyzing quantitative features of calibrated models. Prominent examples in the literature include, among others, Rios-Rull [Rio96] and Heaton and Lucas [HL96].

Unfortunately there are often no theoretical foundations for algorithms that claim to compute competitive equilibria in models with incomplete markets or overlapping generations. In particular, since all computation suffers from truncation and rounding errors, it is obviously impossible to numerically verify that the optimality and market clearing conditions are satisfied, and that a competitive equilibrium is found. The fact that the equilibrium conditions are approximately satisfied generally does not yield any implications on how well the computed solution approximates an exact equilibrium. Computed allocations and prices could be arbitrarily far from competitive equilibrium allocations and prices.

In this paper we develop an error analysis for the computation of competitive equilibria in models with heterogeneous agents where equilibrium prices are infinite dimensional. We define an ε-equilibrium as a collection of finite sets of choices and prices such that there exists a process of prices and choices that takes values exclusively in these sets and for which the relative errors in agents' Euler equations and the errors in market clearing conditions are below some small ε at all times. Existing algorithms for the computation of equilibria in dynamic models can be interpreted as computing ε-equilibria, and the finiteness of ε-equilibria allows us to computationally verify if a candidate solution constitutes an ε-equilibrium. To give an economic interpretation of the concept, we follow Postlewaite and Schmeidler's [PS81] analysis for finite economies and interpret ε-equilibria as approximating exact equilibria of a close-by economy.

In finite economies the problem of interpreting ε-equilibria is easiest illustrated in a standard Arrow–Debreu exchange economy. Scarf [Sca67] proposes a method that approximates equilibria for any given finite economy in the following sense: Given individual endowments ε^i for individuals $i = 1, ..., I$ and an aggregate excess demand function $\xi(p, (e^i)_{i=1}^I)$, and given an $\varepsilon > 0$, the method finds a $\bar{p}$ such that $\|\xi(\bar{p}, (e^i)_{i=1}^I)\| < \varepsilon$. As Anderson (1986) points out, this fact does not imply that it is possible to find a $\tilde{p}$ such that $\|\tilde{p} - p^*\| < \varepsilon$ for some exact equilibrium price vector p^*. Richter and Wong [RW99] make a similar observation. They examine the problem of the computation of equilibria from the viewpoint of computable analysis and point out that while Scarf's algorithm generates a sequence of values converging to a competitive equilibrium, knowing any finite initial sequence might shed no light at all on the limit.

However, if individual endowments are interior and if the value of the excess demand function at $\bar{p}$, $\|\xi(\bar{p}, (e^i))\|$, is small, then $\bar{p}$ is an equilibrium price for a close-by economy. Homogeneity of aggregate excess demand implies that if $\bar{p} \cdot \xi(\bar{p}, (e^i)) = 0$, then $\|(\bar{p}, (e^i)) - (p^*(\tilde{e}^i))\| < \varepsilon$ with $\xi(p^*, (\tilde{e}^i)) = 0$. Figure 1 displays an equilibrium correspondence, which maps endowments into equilibrium prices. The computed price for the original economy is far away from the unique exact equilibrium price. No small perturbation of this price is an equilibrium price for the economy. However, there is an economy with close-by endowments for which the computed price is an equilibrium price.

Researchers rarely know the exact individual endowments of agents anyway, and if close-by specifications of exogenous variables lead to vastly differ-

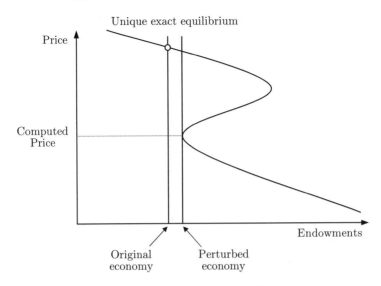

Fig. 1. So close and yet so far

ent equilibria, it will be at least useful to know one possible equilibrium for one realistic specification of endowments. As Postlewaite and Schmeidler [PS81] put it, "If we don't know the characteristics, but rather, we must estimate them, it is clearly too much to hope that the allocation would be Walrasian with respect to the estimated characteristics even if it were Walrasian with respect to the true characteristics."

This issue has long been well understood from the viewpoint of computational mathematics. In general, sources of errors in computations can be classified into three categories. First, there are errors due to the theory: Economic models typically contain many idealizations and simplifications. Second, there are errors due to the specification of exogenous variables: The economic model depends on parameters that are themselves computed approximately and are the results of experimental measurements or the results of statistical procedures. Third, there are errors due to truncation and rounding: Each limiting process must be broken off at some finite stage, and because computers usually use floating point arithmetic, round-off errors result. There exists a debate within the applied economic literature (which uses computations) about the trade-off between the first and third sources of errors, but there is surprisingly little discussion about a possible trade-off between the second and third sources. This paper explores how this latter trade-off can be used to interpret approximate solutions to dynamic general equilibrium models via backward error analysis.

Backward error analysis is a standard tool in numerical analysis that was developed in the late 1950s and 1960s (see Wilkinson [Wil63] or Higham

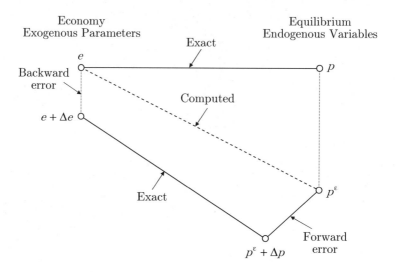

Fig. 2. Mixed forward–backward error analysis

[Hig96]). Surprisingly, this tool has not been widely used in economics.[3] In backward error analysis exogenous parameters are given, an approximate solution is computed, and then the necessary perturbations in exogenous parameters are determined for the computed solution to be exact. The focus of our analysis of popular models in Sections 5 and 6 of this paper is the calculation of backward errors. Due to the nature of economic problems we cannot perform "pure" backward error analysis and only perturb exogenous parameters. Instead, we compute bounds on perturbations of both exogenous parameters and endogenous equilibrium values. Higham [Hig96] calls this "mixed" forward–backward error analysis. Figure 2 elucidates this concept. Ideally we are interested in the exact equilibrium for the original economy and would like to provide a bound on the distance between the computed and the exact equilibrium. However, we argue that this may often be impossible and, instead, we interpret the computed equilibrium as a good approximation (small forward error in the figure) of the exact equilibrium of an economy that is a slight perturbation (backward error) of the original economy.

The analysis in our paper is, from a theoretical perspective, perhaps closest to Mailath, Postlewaite, and Samuelson's [MPS05] discussion of ε-equilibria in dynamic games. An important difference is that Mailath et al. allow for perturbations in the instantaneous payoff functions of the game. In our framework

---

[3] Judd's textbook [Jud98], for example, mentions backward error analysis and provides a citation from the numerical analysis literature, but never applies the concept to an economic problem. Sims [Sim89] and Ingram [Ing90] use the terminology "backsolving" for a method for solving non-linear, stochastic systems. This concept is fairly unrelated to backward error analysis.

this can lead to preferences over payoff streams that are far away from the original preferences. Therefore, we do not consider these.

For models with a single agent, Santos and his coauthors examine forward error bounds both on policy functions and on allocations (Santos and Vigo-Aguiar [SV98], Santos [San00], and Santos and Peralta-Alva [SP05]). They derive sufficient conditions under which it is possible to estimate error bounds from the primitive data of the model and from Euler equation residuals. However, most of these results do not generalize to models with heterogeneous agents and incomplete markets. No sufficient conditions are known that allow the derivation of error bounds on computed equilibrium prices and allocations in the models considered in this paper.

The paper is organized as follows. In Section 2 we illustrate the main intuition in a simple two-period example. Section 3 outlines an abstract dynamic model and defines what we mean by close-by economies. Section 4 develops the theoretical foundations of our method. In Section 5 we apply this method to a model with overlapping generations and production. In Section 6 we apply the methods to a version of Lucas' [Luc78] asset pricing model with heterogeneous agents.

2 The Main Intuition in a Two-period Economy

In this section we demonstrate the main themes of this paper in a simple two-period model. We first show how competitive equilibria can be characterized by a system of equations that relates endogenous variables in one period to endogenous variables of the next period. These equations, which we refer to as the equilibrium equations, enable us later in the paper to describe infinite equilibria with finite sets. Second, we define an ε-equilibrium and provide an example that shows that ε-equilibrium prices and allocations can be a terrible approximation to exact equilibria. We show that, in the example, perturbations in individual endowments can rationalize ε-equilibria as exact equilibria.

Example 1. Consider a simple pure exchange economy with two agents, two time periods, and no uncertainty. There is a single commodity in each period: agents' endowments are (e_0^i, e_1^i) for $i = 1, 2$. Agents can trade a bond that pays one unit in the second period; the price of the bond is denoted by q. Agents' bond holdings are θ^i, $i = 1, 2$. Agents preferences are represented by time-separable utility $U^i(x_0, x_1) = v_i(x_0) + u_i(x_1)$, $i = 1, 2$, for increasing, differentiable, and concave functions $v_i, u_i : \mathbb{R}_+ \to \mathbb{R}$.

A competitive equilibrium is a collection of choices $(c^i, \theta^i)_{i=1,2}$ and a bond price q such that both agents maximize utility and markets clear, i.e., $\theta^1 + \theta^2 = 0$ and for both $i = 1, 2$,

$$(c^i, \theta^i) \in \underset{c \in \mathbb{R}_+^2, \theta \in \mathbb{R}}{\arg\max} \ U^i(c) \quad \text{s.t.} \quad c_0 = e_0^i - q\theta, \ c_1 + e_1^i + \theta.$$

To represent equilibria for infinite-horizon models we want to derive a system of equations that links endogenous variables (i.e., choices and prices) today to endogenous variables in the next period. In this simple example, we define the vector of relevant endogenous variables to consist of current consumption, current portfolios, and current prices, $z = ((c^i, \theta^i)_{i=1,2}, q)$. (Even though agents do not trade the bond in the second period we include zero bond holdings and a zero price in the state variable z_1 for that period. This setup has the advantage that the resulting equilibrium expressions look very similar to those in the infinite-horizon problems that we examine in the main part of the paper.)

In this two-period example, we define a system of equations $h(z_0, \kappa, z_1)$ such that $((\bar{c}^i, \bar{\theta}^i)_{i=1,2}, \bar{q}) \in \mathbb{R}_+^2 \times \mathbb{R}^2 \times \mathbb{R}_+^2 \times \mathbb{R}^2 \times \mathbb{R}_+^2$ is a competitive equilibrium if and only if there exists $\kappa = (\kappa^1, \kappa^2) \in \mathbb{R}_+^2 \times \mathbb{R}_+^2$ such that $h(\bar{z}_0, \kappa, \bar{z}_1) = 0$, with $\bar{z}_0 = ((\bar{c}_0^i, \bar{\theta}_0^i)_{i=1,2}, \bar{q}_0)$ and $\bar{z}_1 = ((\bar{c}_1^i, 0)_{i=1,2}, 0)$. The system is

$$
h(z_0, \kappa, z_1) = \begin{cases}
-q_0 v_i'(c_0^i) + u_i'(c_1^i) - q_0 \kappa_0^i + \kappa_1^i & (i = 1, 2), \\
c_0^i - (e_0^i - q_0 \theta_0^i) & (i = 1, 2), \\
c_1^i - (e_1^i + \theta_0^i) & (i = 1, 2), \\
\kappa_0^i c_0^i & (i = 1, 2), \\
\kappa_1^i c_1^i & (i = 1, 2), \\
\theta_0^1 + \theta_0^2.
\end{cases}
$$

An exact equilibrium is characterized by $h(\cdot) = 0$, but computational methods can rarely find exact solutions. All one can usually hope for is to find an ε-equilibrium, namely (z_0, z_1), such that $\min_{\kappa \in \mathbb{R}_+^4} \|h(z_0, \kappa, z_1)\| < \varepsilon$. Unfortunately, even in this very simple framework, one can construct economies where ε-equilibria can be arbitrarily far from exact equilibria.

2.1 Approximate Equilibria Can Be Far from Exact

Consider the following class of economies parameterized by $\delta > 0$:

$$
v_1(x) = x, \quad u_1(x) = -\frac{1}{x}, \; e^1 = (2, \delta), \quad \text{and}
$$
$$
v_2(x) = -\frac{1}{x}, \; u_2(x) = x, \quad e^2 = (0, 2).
$$

We can easily verify that a competitive equilibrium is given by

$$
q_0 = \frac{1}{(2 + \delta)^2}, \quad \theta_0^1 = 2 = -\theta_0^2,
$$
$$
c^1 = \left(2 - \frac{2}{(2 + \delta)^2}, 2 + \delta \right), \quad c^2 = \left(\frac{2}{(2 + \delta)^2}, 0 \right).
$$

This equilibrium is unique for $\delta > 0$. In addition, for $\delta < 1/\sqrt{4 - \varepsilon} - 1/2$, the following values of the asset price, asset holdings, and consumption vectors yield an ε-equilibrium:

$$q_0 = 4, \quad \theta_0^1 = -\theta_0^2 = \frac{1}{2}, \quad c^1 = \left(0, \frac{1}{2} + \delta\right), \quad c^2 = \left(2, \frac{3}{2}\right).$$

All equations except for $h^1(\cdot) = 0$ for agent 1 hold with equality. The error in this equation is below ε by construction.

This example shows that for any (arbitrarily small) $\varepsilon > 0$ we can construct an economy and an ε-equilibrium that is far from an exact equilibrium both in allocations and prices. Furthermore, it is worth noting that agents' welfare levels differ significantly between the exact equilibrium and the ε-equilibrium. For very small δ, utility levels in the exact equilibrium are approximately $(U^1, U^2) \approx (1, -2)$, while in the ε-equilibrium they are approximately $(U_\varepsilon^1, U_\varepsilon^2) \approx (-2, 1)$. No matter how one looks at it, the ε-equilibrium is evidently a terrible approximation for the exact equilibrium.[4] This observation motivates us to interpret ε-equilibria as approximate equilibria for close-by economies.

2.2 Perturbing Endowments Makes Approximate Equilibria Exact

In the example, we can easily explain the idea of mixed forward–backward error analysis and how an ε-equilibrium can be understood as approximating an exact equilibrium of a close-by economy. At the ε-equilibrium $q_0 = 4$, $\theta_0^1 = -\theta_0^2 = 1/2$, the only equilibrium equation that does not hold with equality is

$$h^1 = -q_0 + \frac{1}{(e_1^1 + \theta_0^1)^2} = -4 + \frac{1}{(e_1^1 + 1/2)^2} = 0.$$

If we replace the endowments e_1^1 by $\hat{e}_1^1 = e_1^1 + w$ for some small w we can evidently set $h^1 = 0$ by using $w = -\delta$. The corresponding perturbation of agent 1's consumption is $c_1^1 = c_1^1 - \delta$. The ε-equilibrium is exact for the perturbed economy. In this mixed forward–backward error analysis we perturbed the value of the endogenous variable c_1^1 and the value of the exogenous parameter e_1^1.

While this is the main idea underlying our error analysis, there is one additional complication that arises when agents live for many periods: Errors may propagate over time, and no sensible bounds on perturbations in endowments can be derived by perturbing endowments every period. In Section 6 we discuss this problem and a possible solution.

---

[4] For finite economies there do exist sufficient conditions that relate approximate equilibria to exact equilibria (see, for example, Blum et al. [BCSS98, ch. 8] and Anderson [And86]). However, these cannot be generalized to the infinite-horizon economies we consider in this paper.

3 A General Model

In this section we fix the main ideas in an abstract framework that encompasses both economies with overlapping generations and economies with infinitely lived agents, as well as economies with and without production. In Sections 5 and 6 we consider two standard models and show how to apply the methods developed in this and the next section.

3.1 The Abstract Economy

Time and uncertainty are represented by a countably infinite tree Σ. Each node of the tree, $\sigma \in \Sigma$, represents a date-event and can be associated with a finite history of exogenous shocks $\sigma = s^t = (s_0, s_1, ..., s_t)$. The process of exogenous shocks (s_t) is a Markov chain with finite support $\mathcal{S} = \{1, ..., S\}$. Given an $S \times S$ transition matrix Π, we define probabilities for each node by $\pi(s_0) = 1$ and $\pi(s^t) = \Pi(s_t|s_{t-1})\pi(s^{t-1})$ for all $t > 1$.

There are L commodities, $\ell \in \mathcal{L}$, at each node. There are countably many individuals $i \in \mathcal{I}$ and countably many firms $k \in K$. An individual $i \in \mathcal{I}$ is characterized by his consumption set X^i, his individual endowments $e^i \in X^i$, his preferences $P^i = \{(x, y) \in X^i \times X^i : x \succ^i y\} \subset X^i \times X^i$, and trading constraints. A firm $k \in K$ is characterized by its production set Y^k. With incomplete financial markets, the objective of the firm is in general not well defined, and there is a large, but inconclusive literature on the subject, following Drèze [Dre74]. In the example below we circumvent the problem by assuming that firms are only active in spot markets. An economy $\mathcal{E}$ is characterized by a demographic structure, assets, technologies and preferences, endowments, and trading constraints. In the concrete models below we describe $\mathcal{E}$ explicitly.

The original economy is assumed to be Markovian. The number of agents active in markets at a given node is finite and time-invariant, but it may depend on the underlying exogenous shock. Agents maximize time and state-separable utility. Firms only make decisions on spot markets. All individual endowments, payoffs of assets, production sets of firms, and spot utility functions of individuals are time-invariant functions of the shock, s, alone.

Perturbations and Backward Errors

For a mixed forward–backward error analysis, we must specify which exogenous parameters can be perturbed and which kind of perturbations are permissible. The exact set of admissible perturbations will be governed by the economic application in mind. It will become clear below that one must always allow for perturbations at all nodes in the event tree and that the resulting economy will no longer be Markovian. We parameterize economies by node-dependent perturbations $w(\sigma) \in \mathcal{W} \subset \mathbb{R}^N$ and write $\mathcal{E}((w(\sigma))_{\sigma \in \Sigma})$ for a given (nonstationary) perturbed economy. In the original economy $w(\sigma) = 0$ for all $\sigma \in \Sigma$. The vector $w(\sigma) = (w^e(\sigma), w^u(\sigma), w^f(\sigma))$ may contain perturbations of endowments, preferences, and production functions. For the error analysis of the dynamic models in Sections 5 and 6 we use the following perturbations.

Endowments: For $\sigma \in \Sigma$, $w^e(\sigma)$ denotes additive perturbations of the endowments of those individuals who are active in markets at node σ. The perturbed individual endowment of an agent i is then $\tilde{e}^i(\sigma) = e^i(\sigma) + w^{ei}(\sigma)$.

Preferences: We assume throughout the paper that preferences can be represented by a time-separable expected utility function. We consider linear additive perturbations to Bernoulli utilities (as is often done in general equilibrium analysis; see, e.g., Mas-Colell [Mas85]). We assume that for an infinitely lived agent i there exists a strictly increasing, strictly concave, and differentiable Bernoulli function $u^i : \mathbb{R}^L_{++} \times \mathcal{S} \to \mathbb{R}$ such that

$$U^i(x) = \sum_{t=0}^{\infty} \beta^t \sum_{s^t} \pi(s^t) u^i(x(s^t), s_t).$$

Agents have common beliefs Π and discount factors β, and in the original unperturbed economy, Bernoulli utilities only depend on the current shock. Given $u^i(x, s_t)$ and a utility perturbation $w^{ui}(s^t) \in \mathbb{R}^L$, the perturbed Bernoulli utility is

$$\tilde{u}^i(x, s^t) = u^i(x, s^t) + w^{ui}(s^t) \cdot x.$$

These perturbations are difficult to interpret economically since they are not invariant under affine transformations of the original Bernoulli function, but using such linear utility perturbations simplifies the exposition and the notation. We show below how to compute economically meaningful error bounds from these perturbations and properties of the Bernoulli function u^i.

Production Functions: We assume in Section 5 that at each node s^t there is an aggregate technology described by a production function $f_{s_t} : \mathbb{R}^{\ell_1} \to \mathbb{R}$. We consider linear perturbations and write at node s^t,

$$\tilde{f}_{s_t}(y(s^t)) = f_{s_t}(y(s^t)) + w^f(s^t) \cdot y(s^t).$$

Close-By Economic Agents

It is crucial for the analysis that the suggested perturbations of exogenous parameters result in economies that are close-by to the original economy in a meaningful way. The first step in our argument is to define an appropriate topology on the space of possibly perturbed economies. We choose the supnorm to measure the size of perturbations along the event tree, since we want the perturbed economies to stay as close by as possible to the original economy. Throughout the paper, for a vector $x \in \mathbb{R}^n$, $\|x\|$ denotes the sup-norm, $\|x\| = \max\{|x_1|, ..., |x_n|\}$.

Define the space of perturbations to be

$$\ell_\infty(\Sigma, \mathcal{W}) = \left\{ w = (w^e(\sigma), w^u(\sigma), w^f(\sigma)) : \sup_{(\sigma, w) \in \Sigma \times W} \|w(\sigma)\| < \infty \right\},$$

with $\|x\| = \sup_{\sigma \in \Sigma} \|x(\sigma)\|$ for a sequence $x \in \ell_\infty$. Naturally, a perturbed economy is close to the original economy if the sup-norm of the perturbations is small. For the endowments in the perturbed economy, this obviously means that they are close to the original endowments at all nodes of the tree.

While small differences in individual endowments are easy to understand, differences in utility functions and production functions are more difficult to interpret. In particular, it is obviously sensible to think of utility perturbations in terms of the implied difference in agents' underlying preferences (and not in "utils" as implied by the linear additive perturbations). Following Postlewaite and Schmeidler [PS81] and Debreu [Deb69] we use the Hausdorff distance between sets, d^H, to quantify closeness of two preferences P and P'. The distance between two preferences is

$$d^H(P, P') = \max \left\{ \sup_{(x,y) \in P} \left(\inf_{(x',y') \in P'} \|(x,y) - (x' - y')\| \right), \right.$$
$$\left. \sup_{(x',y') \in P'} \left(\inf_{(x,y) \in P} \|(x,y) - (x',y')\| \right) \right\}.$$

Linearly perturbed utility

$$\tilde{U}^i(x) = \sum_{t=0}^\infty \beta^t \sum_{s^t} \pi(s^t)(u^i(x, s^t) + w^{ui}(s^t) \cdot x)$$

generally does not represent preferences that are close to the original preferences. However, the following lemma implies that if one finds an exact equilibrium for an economy with utility functions $(\tilde{U}^i)$, there also exists an economy with individual preferences close by to the original preferences for which the same prices and allocation constitute a competitive equilibrium. For simplicity, we assume in the lemma that preferences are homothetic. This allows us to derive explicit bounds on the distance between perturbed preferences and the original preferences.

Lemma 1. *Suppose that the time-separable expected utility function U represents homothetic preferences P. Given choices $\bar{x} \in \ell_\infty(\Sigma, \mathcal{L})$ and perturbations $w \in \ell_\infty$, suppose some $\delta \in \ell_\infty$ satisfies $D_x u(\delta_1(s^t)\bar{x}_1(s^t), ..., \delta_L(s^t)\bar{x}_L(s^t), s_t) = D_x u(\bar{x}(s^t), s_t) + w(s^t)$ for all s^t. If*

$$\bar{x} \in \arg\max \tilde{U}(x) \quad \text{s.t.} \quad x \in \mathcal{B},$$

for some convex set $\mathcal{B} \subset \ell_\infty$ with $0 \in \mathcal{B}$, then there exist convex and increasing preferences P' such that whenever $(y, \bar{x}) \in P'$, then $y \notin \mathcal{B}$, i.e., $\bar{x}$ is the best choice in $\mathcal{B}$, and

$$d^H(P, P') \leq \sup_{s^t, \ell} \bar{x}_\ell(s^t) \left(1 - \frac{\delta_\ell(s^t)}{\sup_\sigma \|\delta(\sigma)\|} \right).$$

The lemma, which is proven in the Appendix, shows that for given perturbations to marginal utilities one can construct close-by preferences that support the desired choices. It also shows how to bound the distance between the original preferences and the perturbed preferences. Although the discussion of the perturbations in Section 4 is presented in terms of linear utility perturbations, the reader should keep in mind that such perturbations can be translated to differences in the underlying preferences, which is an economically more meaningful measure.

In applications, it is often tempting to perturb conditional probabilities and node-dependent discount factors. However, such perturbations may lead to preferences that are very far away from the original preferences in our norm. Perturbations in resulting unconditional probabilities may get arbitrarily large for date-events far along the event tree, and therefore marginal rates of substitution for the perturbed preferences will be far from those of the original preferences. The preferences will be far in the Hausdorff distance. For this reason, these perturbations will not be considered in this paper.

3.2 Equilibrium Equations

A competitive equilibrium for the economy $\mathcal{E}((w(\sigma))_{\sigma \in \Sigma}$ is a process of endogenous variables $(z(\sigma))_{\sigma \in \Sigma}$ with $z(\sigma) \in \mathcal{Z} \subset \mathbb{R}^M$, which solve agents' optimization problems and clear markets. The set $\mathcal{Z}$ denotes the set of all possible values of the endogenous variables. We refer to the collection of the economy and the endogenous variables, $(\mathcal{E}((w(\sigma))_{\sigma \in \Sigma}), (z(\sigma))_{\sigma \in \Sigma})$, as an *economy in equilibrium*.

In many dynamic economic models an equilibrium can be characterized by a set of equations that relates current-period exogenous and endogenous variables to endogenous variables one period ahead, as well as by a set of equations that restricts current endogenous variables to be consistent with feasibility and optimality. Examples of such conditions are individuals' Euler equations, firms' first-order conditions, and market clearing equations for goods or financial assets. For our error analysis we assume that such a set of equations is given and denote it by

$$h(\bar{s}, \bar{z}, \bar{w}, \kappa, z_1, ..., z_S) = 0.$$

The arguments $(\bar{s}, \bar{z}, \bar{w})$ denote the exogenous shock, the endogenous variables, and the perturbations for the current period. For each subsequent exogenous shock s, z_s denotes endogenous variables. The variables $\kappa \in \mathcal{K}$ should be thought of as representing Kuhn–Tucker multipliers or slack variables in inequalities. These variables are used to transform inequalities into equations.

Throughout the analysis, we impose assumptions on preferences and technology, ensuring that $(\mathcal{E}((w(\sigma))_{\sigma \in \Sigma}), (z(\sigma))_{\sigma \in \Sigma})$ is an economy in equilibrium if and only if for all $s^t \in \Sigma$, there exist $\kappa(s^t) \in \mathcal{K}$ such that

$$h(s_t, z(s^t), w(s^t), \kappa(s^t), z(s^t 1), ..., z(s^t S)) = 0.$$

We refer to $h(\cdot) = 0$ as the equilibrium equations.

4 Approximate Equilibria and Their Interpretation

The applied computational literature usually refers to recursive equilibria. A recursivity assumption is crucial for computational tractability, because one must find a simple way to represent infinite sequences of allocations and prices. These equilibria are characterized by policy functions that map the current "state" of the economy into choices and prices, and by transition functions that map the state today into a probability distribution over the next period's state. Unfortunately, in dynamic GEI models, recursive equilibria do not always exist, and no nontrivial assumptions are known that guarantee the existence of recursive equilibria (for counterexamples to existence, see, e.g., Hellwig [Hel82], Kubler and Schmedders [KS02], and Kubler and Polemarchakis [KP04]). Therefore, one cannot evaluate the quality of a candidate equilibrium by a calculation of how close the computed policy function is to an exact policy function. Instead, we need to define a notion of approximate equilibrium that is general enough to exist in most interesting specifications of the model and is also tractable in the sense that actual approximations in the literature can be interpreted as such equilibria (or at least that these equilibria can be constructed fairly easily from the output of commonly used algorithms). In most popular models, recursive ε-equilibria exist (even if exact recursive equilibria fail to exist). We therefore build our error analysis on the concept of recursive ε-equilibrium.

The relevant endogenous state space $\Psi \subset \mathbb{R}^D$ depends on the underlying model and is determined by the payoff-relevant, predetermined endogenous variables; that is, by variables sufficient for the optimization of individuals at every date-event, given the prices. The value of the state variables $(s_0, \psi_0) \in \mathcal{S} \times \Psi$ in period 0 is called initial condition and is part of the description of the economy. A recursive ε-equilibrium is defined as follows.

Definition 1. A *recursive ε-equilibrium* consists of a finite state space Ψ, a policy function $\rho : \mathcal{S} \times \Psi \to \mathbb{R}^{M-D}$, as well as transition functions $\tau_{ss'} : \Psi \to \Psi$, for all $s, s' \in \mathcal{S}$, such that for all states $(\bar{s}, \bar{\psi}) \in \mathcal{S} \times \Psi$, the errors in equilibrium equations at the values implied by policy and transition functions, i.e., at $\bar{z} = (\bar{\psi}, \rho(\bar{s}, \bar{\psi})) \in \mathcal{Z}$ and $(z_1, ..., z_S)$ with $z_s = (\psi_s, \rho(s, \psi_s))$, $\psi_s = \tau_{\bar{s}s}(\bar{\psi})$ for all $s \in \mathcal{S}$, are below ε. That is, they satisfy $\min_{\kappa \in \mathcal{K}} \|h(\bar{s}, \bar{z}, \mathbf{0}, \kappa, z_1, ..., z_S)\| < \varepsilon$.

A recursive ε-equilibrium consists of a discretized state space as well as of policy and transition functions, which imply that errors in equilibrium equations are always below ε. In most contexts it will be straightforward to derive the transition function from the policy function. For example, in a finance economy, the beginning-of-period portfolio holdings constitute the endogenous state. The policy function assigns new portfolio holdings, which then form the endogenous state next period. In these cases the recursive ε-equilibrium is completely characterized by the policy function.

We define the state space as a finite collection of points in order to verify whether a candidate solution constitutes an ε-equilibrium. With this definition, the verification involves checking only a finite number of inequalities. Some popular recursive methods (see Judd [Jud98]) rely on smooth approximations of policy and transition functions using orthogonal polynomials or splines, but we can always extract a recursive ε-equilibrium with a discretized state space from such smooth approximations.

Recursive methods enable us to approximate an infinite-dimensional equilibrium by a finite set. Given an initial value of the shock, s_0, and initial values for the endogenous state, ψ_0, a recursive ε-equilibrium assigns a value of endogenous variables to any node in the infinite event tree: For any node s^t, the value of the endogenous state is given by $\psi(s^t) = \tau_{s_{t-1}s_t}(\psi(s^{t-1}))$, and the value of the other endogenous variables is given by $\rho(s_t, \psi(s^t))$. We call the resulting stochastic process an ε-equilibrium process and write $(z^\varepsilon(\sigma))_{\sigma \in \Sigma}$. It will be useful to define ε-equilibrium sets, $\mathcal{F} = (\mathcal{F}_1, ..., \mathcal{F}_S)$ to be the graph of the policy function, so $\mathcal{F}_s \subset \mathcal{Z}$, $\mathcal{F}_s = \mathrm{graph}(\rho(s, \cdot))$.

4.1 Error Analysis

Judd [Jud92] and den Haan and Marcet [DM94] suggest evaluating the quality of a candidate solution by using Euler equation residuals. In these methods, relative maximal errors in Euler equations of a usually imply that the solution describes a recursive ε-equilibrium. Unfortunately, the example in Section 2 shows that the computed ε-equilibrium may be far away from an exact equilibrium for the economy, no matter how small ε is. In other words, we cannot perform a pure forward error analysis. As a consequence we perform a mixed forward–backward error analysis and interpret ε-equilibria as approximations to exact equilibria of close-by economies. In infinite-horizon models, the question now becomes, what is meant by an approximation to an exact equilibrium?

Ideally a recursive ε-equilibrium would generate an ε-equilibrium process that is close by to a competitive equilibrium for a close-by economy at all date events. If this were the case, one could find small perturbations of exogenous parameters such as endowments and preferences of the original economy so that the perturbed economy has a competitive equilibrium that is well approximated by the ε-equilibrium process at *each* node of the event tree. This idea is formalized in the following definition of approximate equilibrium.

Definition 2. Given an economy $\mathcal{E}$, an ε-equilibrium process $(z^\varepsilon(\sigma))_{\sigma \in \Sigma}$ is called *path-approximating* with error δ if there exists an economy in equilibrium, $(\mathcal{E}((w(\sigma))_{\sigma \in \Sigma}), (\hat{z}(\sigma))_{\sigma \in \Sigma}$ with $\sup_{\sigma \in \Sigma} \|w(\sigma)\| < \delta$ and $\sup_{\sigma \in \Sigma} \|z^\varepsilon(\delta) - \hat{z}(\sigma)\| < 0$.

In models with finitely lived agents, e.g., OLG models, ε-equilibria will usually be path-approximating. However, in models where agents are infinitely lived, we cannot expect that a recursive ε-equilibrium gives rise to a process that path-approximates a close-by economy in equilibrium. If agents make small errors in their choices each period, these errors are likely to propagate over time, and after sufficiently many periods the ε-equilibrium allocation will be far away from the exact equilibrium allocation. The following simple example illustrates the ease of constructing ε-equilibria that do not path-approximate an economy in equilibrium, no matter how small ε is.

Example 2. Consider an infinite-horizon exchange economy with two infinitely lived agents, a single commodity, and no uncertainty. Suppose that agents have identical initial endowments $e^1 = e^2 > 0$ in each period and identical preferences with $u_i(c_t) = \log(c_t)$ and with a common discount factor $\beta \in (0,1)$. There is a consol in unit net supply that pays 1 unit of the consumption good each period. The price of the consol is q_t, portfolios are θ_t^i. Each agent i, $i = 1, 2$, faces a short-sale constraint $\theta_t^i \geq 0$ for all t.

Let the endogenous variables be $z = ((\theta_-^i, \theta^i, c^i, m^i))_{i=1,2}, q)$. Admissible perturbations are $w = (w^{ui}, w^{ei})_{i=1,2} \in \mathbb{R}^4$, so we allow for perturbations both in endowments and in preferences. The equilibrium equations $h(\bar{z}, \bar{w}, \kappa, z) = 0$ with $h = (h^1, ..., h^6)$ are

$$h^1 = -1 + \beta \frac{(1+q)m^i}{\bar{q}\bar{m}^i} + \kappa^i \qquad (i = 1, 2),$$
$$h^2 = \kappa^i \bar{\theta}^i \qquad\qquad\qquad (i = 1, 2),$$
$$h^3 = \bar{c}^i - \bar{\theta}_-^i (\bar{q} + 1) + \bar{\theta}^i \bar{q} - (e^i + \bar{w}^{ei}) \quad (i = 1, 2),$$
$$h^4 = \theta_-^i - \bar{\theta}^i \qquad\qquad\qquad (i = 1, 2),$$
$$h^5 = \bar{m}^i - (u_i'(\bar{c}^i) + \bar{w}^{ui}) \qquad (i = 1, 2),$$
$$h^6 = \bar{\theta}^1 + \bar{\theta}^2 - 1.$$

The natural endogenous state space for this economy consists of beginning-ofperiod consol holdings. We build market clearing into the state space and only consider θ_-^1, θ_-^2 with $\theta_-^1 + \theta_-^2 = 1$. We write $\psi = \theta_-^1$ to represent a typical state of the economy, implicitly assuming market clearing. For any initial condition $\psi_0 \in (0, 1)$ the unique exact equilibrium is no trade in the consol, with each agent consuming $\theta_{0-}^i + e^i$ every period, where $\theta_{0-}^1 = \psi_0 = 1 - \theta_{0-}^2$. The consol price is $q_t = \beta/(1 - \beta)$ for all $t \geq 0$.

Now suppose each period agent 1 sells a small amount of the consol to agent 2. As a result, agent 1's consumption converges to e^1 while the consumption of agent 2 converges to $e^2 + 1$. There is no economy with close-by endowments for which this allocation is an approximate equilibrium allocation. We can construct a recursive ε-equilibrium as follows. Define

$$\tau(\psi) = \begin{cases} \psi - \delta, & \text{if } \psi > \delta, \\ \psi, & \text{otherwise.} \end{cases}$$

Define $q = \rho_q(\psi) = \beta/(1-\beta)$, $\theta^1 = \rho_{\theta^1}(\psi) = \tau(\psi)$, and $c^1(\psi) = e^1 + \psi(q + 1) - \theta q$, $c^2(\theta) = 1 + e^1 + e^2 - c^1(\theta)$. These functions describe a recursive ε-equilibrium as long as $0 \leq \delta \leq \varepsilon(1-\beta)e^1$. Except for the Euler equations h^1, all equilibrium equations hold with equality. For $\theta_- = \psi > 2\delta$, the error in Euler equations for agent 1 is given by

$$\|h^1\| = \left| -1 + \frac{e^1 + (\theta + \delta)(q+1) - \theta q}{e^1 + \theta(q+1) - \theta q + \delta q} \right| < \frac{\delta(q+1)}{e^1}.$$

For $2\delta \geq \psi > \delta$, we have

$$\|h^1\| = \left| -1 + \frac{e^1 + (\theta + \delta)(q+1) - \theta q}{e^1 + \theta(q+1) - \theta q} \right| < \frac{\delta(q+1)}{e^1}.$$

and finally for $\psi < \delta$, we have $h^1 = 0$. The argument for agent 2 is analogous. For the initial condition $\psi_0 = 0.5$, the constructed recursive equilibrium yields an ε-equilibrium process which, in the sup-norm, is far from any exact equilibrium of a close-by economy. Figure 3 shows the exact equilibrium and the ε-equilibrium process.

Note that this phenomenon is a general problem that does not only occur in economies with incomplete markets. The same phenomenon can even arise for an approximate solution to a single-agent decision problem. In the applied literature this problem is commonly addressed by using a weaker notion of approximate equilibrium:[5] A computed solution is considered a good approximation if the computed policy function is close by the true policy function.

We generalize this idea and apply it to our general framework. Instead of requiring that the exact equilibrium process is well approximated by the ε-equilibrium process, we merely require that it is well approximated by the ε-equilibrium set: For each node $s^t \in \Sigma$, given the value of the endogenous state of the exact equilibrium process, there is a state close by such that the value of the (ε-equilibrium) policy function at this state is close to the value of endogenous variables of the exact equilibrium. The following definition formalizes this weaker notion of approximation.

Definition 3. A recursive ε-equilibrium with equilibrium set $\mathcal{F}$ for the economy $\mathcal{E}$ is called *weakly approximating* with error δ if there exists an economy in equilibrium $(\mathcal{E}((w(\sigma))_{\sigma \in \Sigma}), (\hat{z}(\sigma))_{\sigma \in \Sigma})$, with $\sup_{\sigma \in \Sigma} \|w(\sigma)\| < \delta$ such that for all $s^t \in \Sigma$ and all $\hat{z}(s^t)$ there exists a $z \in \mathcal{F}_{s_t}$ which satisfies $\|z - \hat{z}(s^t)\| < \delta$.

---

[5] A notable exception is Santos and Peralta-Alva [SP05] who derive sufficient conditions for sample-path stability in a representative agent model.

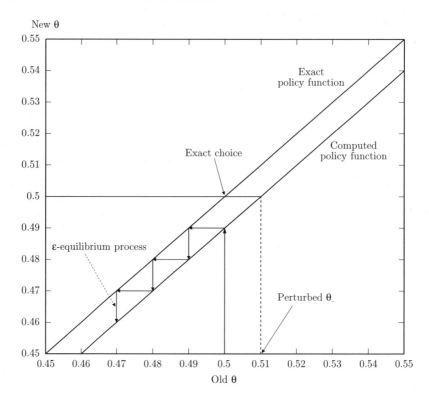

Fig. 3. Weak approximation

Intuitively, the definition requires that for the recursive ε-equilibrium the policy function is close to the policy function of an exact recursive equilibrium (of a close-by economy). In the models we consider in this paper existence of exact recursive equilibria cannot be established. Therefore, we state the definition in terms of competitive equilibria $\hat{z}(\sigma)_{\sigma \in \Sigma}$ of the close-by economy $\mathcal{E}((w(\sigma))_{\sigma \in \Sigma})$.

This condition is much weaker than requiring that the ε-equilibrium process path-approximates an economy in equilibrium. This is to be expected, since closeness in policy functions generally does not yield any implications about how close equilibrium allocations are, even in models where recursive equilibria do exist. The definition only requires that there exists some process with values in $\mathcal{F}$ that approximates the exact equilibrium but does not explicitly state how to construct this process.

The ε-equilibrium in Example 2 weakly approximates the exact equilibrium. For any initial condition the exact equilibrium (for an economy that does not need to be perturbed) involves no trade and an asset price of $\beta/(1 - \beta)$. For the same initial condition the recursive ε-equilibrium implies the same asset price and trading of less than δ units of the asset.

In general, of course, verifying that an ε-equilibrium satisfies Definition 3 will not be as straightforward as in this example, because the exact equilibrium is not known. We explain this problem more carefully with a concrete example in Section 6. The strategy there is illustrated in Figure 3. Given an approximate policy function and an initial state (in the figure, $\theta_- = 0.5$), we try to find a point in the state space where the value of the approximate policy function equals the value of the exact policy function at the initial state.

5 A Model with Overlapping Generations and Production

As a first application of our methods we consider a model of a production economy with overlapping generations. In this model the ε-equilibrium process path-approximates an economy in equilibrium and we derive bounds on the distance between the close-by economy and the specified economy.

5.1 The Economy

Recall that each node of the event tree represents a history of exogenous shocks to the economy, $\sigma = s^t = (s_0, s_1, ..., s_t)$. The shocks follow a Markov chain with finite support $\mathcal{S}$ and with transition matrix Π. Three commodities are traded at each date event: labor, a consumption good, and a capital good that can only be used as input to production and yields no utility. The economy is populated by overlapping generations of agents that live for $N + 1$ periods, $a = 0, ..., N$. An agent is fully characterized by the node in which she is born (s^t). When there is no ambiguity we index the agent by the date of birth. An agent born at node st has nonnegative labor endowment over her life cycle, which depends on the exogenous shock and age, $\mathbf{e}^a(s)$, for ages $a = 0, ..., N$ and shocks $s \in \mathcal{S}$. Agents have no endowments in the capital and the consumption good. The price of the consumption good at each date event is normalized to 1, the price of capital is denoted by $p_k(s^t)$, and the market wage is $p_\ell(s^t)$. The agent has an intertemporal, von Neumann–Morgenstern utility function over consumption, c, and leisure, ℓ, over his life cycle,

$$U^{s^t} = E_{s^t} \sum_{a=0}^{N} \beta^a u\left(c(s^{t+a}), \ell(s^{t+a}); s_{t+a}\right).$$

The Bernoulli utility u may depend on the current shock s and is strictly increasing, strictly concave, and continuously differentiable. We denote the partial derivatives by u_c and u_ℓ.

Households have access to a storage technology. They can use one unit of the consumption good to obtain one unit of the capital good in the next period. We denote the investment of household t at node s^{t+a} into this technology by $\theta^t(s^{t+a})$. All agents are born with zero assets, $\theta^t(s^{t-1}) = 0$. We do not restrict investments to be nonnegative, thus allowing households to borrow against future labor income.

There is a single representative firm, which in each period uses labor and capital to produce the consumption good according to a constant returns to scale production function $f(K, L; s)$, given shock $s \in S$. Firms make decisions on how much capital to buy and how much labor to hire after the realization of the shock s_t, and so they face no uncertainty and simply maximize current profits. At time t the household sells all its capital goods accumulated from last period, $\theta^\sigma(s^{t-1})$, to the firm for a market price $p_k(s^t) > 0$.

For given initial conditions s_0, $((\theta^t(s_{-1}))_{t=-N}^0$ a competitive equilibrium is a collection of choices for households $(c^t(s^{t+a}), \ell^t(s^{t+a}), \theta^t(s^{t+a}))_{a=0}^N$, for the representative firm $(K(s^t), L(s^t))$, as well as prices $(p_k(s^t), p_\ell(s^t))$ such that households and the firm maximize and markets clear for all s^t.

We want to characterize competitive equilibria by equilibrium equations. We define the endogenous variables at some node to consist of individuals' investments from the previous period, $\theta_- = (\theta_-^0, ..., \theta_-^N)$, new investments, $\theta = (\theta^0, ..., \theta^N)$, consumption and leisure choices, $c = (c^0, ..., c^N)$ and $\ell = (\ell^0, ..., \ell^N)$ as well as Lagrange multipliers $\lambda = (\lambda^0, ..., \lambda^N)$, the firm's choices K, L, and spot prices, (p_ℓ, p_k), so $z = (\theta_-, \theta, c, \ell, \lambda, K, L, p_k, p_\ell)$. We build bounds and normalizations into the admissible endogenous variables, i.e., we only consider z for which $\theta_-^0 = 0$, $\theta^N = 0$, $c, \ell, \lambda \geq 0$, $K, L \geq 0$. We consider perturbations in individual endowments, in preferences and in production functions, i.e., define $w(\sigma) = (w^e(\sigma), w^u(\sigma), w^f(\sigma)) \in \mathbb{R}^{N+1} \times \mathbb{R}^{2(N+1)} \times \mathbb{R}^2$ to be perturbations in endowments and preferences across all agents alive and in the production function at node σ. We write the perturbed Bernoulli function of an agent of age a as

$$u(c, \ell, s) + w^{ua} \cdot \begin{pmatrix} c \\ \ell \end{pmatrix} \quad \text{for} \quad w^{ua} \in \mathbb{R}^2.$$

Production functions are perturbed in a similar fashion:

$$f(K, L, s) + w^f \cdot \begin{pmatrix} K \\ L \end{pmatrix}.$$

Equilibrium is characterized by equilibrium equations; $\bar{s}$, $\bar{z}$, $\bar{w}$, and $z(1), ..., z(S)$ are consistent with equilibrium if and only if $h(\bar{s}, \bar{z}, \bar{w}, z(1), ..., z(S)) = 0$ with

$$
h = \begin{cases}
\theta^a_-(s) - \bar{\theta}^{a-1} & (a = 1, ..., N, \ s \in \mathcal{S}) & (h^1), \\
-\bar{\lambda}^{a-1} + \beta \sum_{s \in \mathcal{S}} \Pi(s|\bar{s}) \lambda^a(s) p_k(s) & (a = 1, ..., N) & (h^2), \\
\bar{\theta}^a \bar{p}_k + (e^a(\bar{s}) + \bar{w}^{ea} - \bar{\ell}^a) \bar{p}_\ell - \bar{\theta}^a - \bar{c}^a & (a = 1, ..., N) & (h^3), \\
u_c(\bar{c}^a, \bar{\ell}^a, \bar{s}) + \bar{w}^{ua}_1 - \bar{\lambda}^a & (a = 1, ..., N) & (h^4), \\
u_\ell(\bar{c}^a, \bar{\ell}^a, \bar{s}) + \bar{w}^{ua}_2 - \bar{\lambda}^a \bar{p}_\ell & (a = 1, ..., N) & (h^5), \\
\bar{p}_k - (f_K(\bar{K}, \bar{L}, \bar{s}) + \bar{w}^f_1) & & (h^6), \\
\bar{p}_\ell - (F_L(\bar{K}, \bar{L}, \bar{s}) + \bar{w}^f_2) & & (h^7), \\
\sum_{a=0}^{N} \bar{\theta}^a - \bar{K} & & (h^8), \\
\sum_{a=0}^{N} (e^a(\bar{s}) + \bar{w}^{ea} - \bar{\ell}^a) - \bar{L} & & (h^9).
\end{cases}
$$

We denote equation $h^i(\cdot) = 0$ by $(h_{\mathrm{olg}}i)$ for $i = 1, ..., 9$. Under standard assumptions on preferences and the production function, which guarantee that first-order conditions are necessary and sufficient, a competitive equilibrium can be characterized by these equations. The natural endogenous state space for a recursive equilibrium consists of individuals' beginning-of-period holdings in the capital good θ_-. Kubler and Polemarchakis [KP04] prove the existence of recursive ε-equilibria. The equilibrium values of all variables are given by a policy function ρ. For example, we write $\rho_K(s, \theta_-)$ for the policy term that determines aggregate capital K. For a given recursive ε-equilibrium, the transition function is given by equation $(h_{\mathrm{olg}}1)$ which we assume to hold exactly, i.e., $\theta^a_-(s) = \bar{\theta}^{a-1}$ for all a, s. These functions then also determine an ε-equilibrium process.

5.2 Error Analysis

Given an ε-equilibrium $\mathcal{F}$, the objective of our error analysis is to provide uniform bounds on necessary perturbations of the underlying economy and on the distance of the ε-equilibrium process $(z^\varepsilon(\sigma))$ from the exact equilibrium of the perturbed economy. We first perturb only preferences, i.e., we construct a close-by economy with $w^e = w^f = 0$, but allow wu to be nonzero, and show that the ε-equilibrium process is close to the exact equilibrium of this economy. Subsequently we outline one different set of perturbations. We show that it suffices to perturb technology and endowments, holding preferences fixed.

We show how to construct bounds on forward errors, Δ^F, and bounds on backward errors, Δ^B, such that there exists a close-by economy with an exact equilibrium $(\hat{z}(\sigma))_{\sigma \in \Sigma}$ that satisfies $\|\hat{z}(\sigma) - z^\varepsilon(\sigma)\| < \Delta^F$ and $\|w(\sigma)\| < \Delta^B$ for all $\sigma \in \Sigma$, including

$$
\hat{\theta}_-(\sigma) = \theta^\varepsilon_-(\sigma), \ \hat{\theta}(\sigma) = \theta^\varepsilon(\sigma), \ \hat{\ell}(\sigma) = \ell^\varepsilon(\sigma),
$$
$$
\hat{K}(\sigma) = K^\varepsilon(\sigma), \ \hat{L}(\sigma) = L^\varepsilon(\sigma).
$$

By allowing for backward errors, only prices, consumption values, and Lagrange multipliers need to be perturbed. All other endogenous variables in the exact equilibrium $(\hat{z}(\sigma))_{\sigma \in \Sigma}$ take the corresponding values from the ε-equilibrium process. Intuitively, our perturbation procedure is best described as a "backward substitution" approach. For every state $(\bar{s}, \bar{z})$ with $\bar{z} \in \mathcal{F}_{\bar{s}}$ we reduce the errors equation by equation to zero by perturbing some endogenous variables. Of course, we have to keep track of how perturbations in one equation affect the errors in other equations where perturbed variables also appear. The possibility of using backward errors is crucial to ensure that this backward substitution approach successfully sets all equilibrium equations simultaneously to zero. We now provide the technical details.

Consider a particular $\bar{s}, \bar{z} \in \mathcal{F}_{\bar{s}}$ and values of next period variables $(z(1), ..., z(S)) \in \mathcal{F}$, where

$$z(s) = \left(\bar{\theta}, \rho_\theta(s, \bar{\theta}), \rho_c(s, \bar{\theta}), \rho_\ell(s, \bar{\theta}), \rho_\lambda(s, \bar{\theta}),\right.$$
$$\left. \rho_K(s, \bar{\theta}), \rho_L(s, \bar{\theta}), \rho_{p_k}(s, \bar{\theta}), \rho_{p_\ell}(s, \bar{\theta})\right).$$

By definition, these points satisfy $\|h(\bar{s}, \bar{z}, 0, z(1), ..., z(S))\| < \varepsilon$. We assume, without loss of generality, that for all $\bar{s} \in \mathcal{S}$ and all $\bar{z} \in \mathcal{F}_{\bar{s}}$, $\sum_{a=0}^{N}(\mathbf{e}^a(\bar{s}) - \bar{\ell}^a) = \bar{L}$ and $\sum_{a=0}^{N} \bar{\theta}^a - \bar{K} = 0$. Both conditions can be built into the construction of $\bar{z}$. Consequently, the equations given by $(h_{\mathrm{olg}}8)$ and $(h_{\mathrm{olg}}9)$ hold exactly.

To set other equations exactly equal to zero, we must perturb values in $\bar{z}$, which in turn affect other equations. We keep track of the possibly increased errors in the other equations and adjust them accordingly. Moreover, we perturb $\lambda(s)$, $s = 1, ..., S$, for equality in $(h_{\mathrm{olg}}2)$. Such perturbations then fix $\lambda(s)$ and therefore enter the Euler equations $(h_{\mathrm{olg}}2)$ and equations $(h_{\mathrm{olg}}4)$ and $(h_{\mathrm{olg}}5)$ for the subsequent period. As a result, we must be careful how we per-form our perturbation analysis to set those equations equal to zero.

We adjust prices in order to set equations $(h_{\mathrm{olg}}6)$ and $(h_{\mathrm{olg}}7)$ to zero. Maximal errors in prices are simply given by

$$\Delta_1 = \max_{\bar{s} \in \mathcal{S}, \bar{z} \in \mathcal{F}_{\bar{s}}} \|\bar{p}_k - f_K(\bar{K}, \bar{L}, \bar{s})\| \quad \text{and}$$
$$\Delta_2 = \max_{\bar{s} \in \mathcal{S}, \bar{z} \in \mathcal{F}_{\bar{s}}} \|\bar{p}_\ell - f_L(\bar{K}, \bar{L}, \bar{s})\|.$$

For equations $(h_{\mathrm{olg}}3)$ to hold exactly, individual consumption values $\bar{c}^a$ must be perturbed. Given the corrected prices, maximal necessary perturbations in individuals' consumptions are then

$$\Delta_3 = \max_{\bar{s} \in \mathcal{S}, \bar{z} \in \mathcal{F}_{\bar{s}}, a=0,...,N} \left\| \bar{\theta}^a_- f_K(\bar{K}, \bar{L}, \bar{s})\right.$$
$$\left. + (\mathbf{e}^a(\bar{s}) - \bar{\ell}^a) f_L(\bar{K}, \bar{L}, \bar{s}) - \bar{\theta}^a - \bar{c}^a \right\|.$$

In practice the perturbations in prices and consumptions will be tiny since the respective equations can be solved very precisely. More significant errors may arise, however, through cumulative perturbations in λ over an agent's life-span. Equations $(h_{\text{olg}}2)$ show that if the current λ has been perturbed away from $\bar{\lambda}$ for last period's Euler equation, the necessary perturbations in $\lambda(1), ..., \lambda(S)$ might propagate over time. To analyze this problem, define

$$\delta_2^a = \max_{\bar{s}, \bar{z} \in \mathcal{F}_{\bar{s}}} \left\| -\bar{\lambda}^{a-1} + \beta \sum_{s \in \mathcal{S}} \Pi(s, \bar{s}) \rho_{\lambda^a}(s, \bar{\theta}) f_K(\rho_K(s, \bar{\theta}), \rho_L(s, \bar{\theta}), s) \right\|,$$

where the prices have already been replaced by their perturbed values. Note that δ_2^a is an upper bound on errors in equations $(h_{\text{olg}}2)$ only if $\tilde{\lambda} - \bar{\lambda}$. For the general case where $|\tilde{\lambda} - \bar{\lambda}| < \delta_1$, the maximal error is bounded by $\delta_1 + \delta_2^a$, by the triangle inequality.

The perturbations of Lagrange multipliers are mirrored by perturbations of marginal utilities in order to ensure that equations $(h_{\text{olg}}4)$ and $(h_{\text{olg}}5)$ hold exactly. Define

$$\Delta_4 = \max_{\bar{s} \in \mathcal{S}, \bar{z} \in \mathcal{F}_{\bar{s}}, a=0,...,N} \| u_c(\bar{\theta}^a_- f_K(\bar{K}, \bar{L}, \bar{s}) + (\mathbf{e}^a(\bar{s}) - \bar{\ell}^a) f_L(\bar{K}, \bar{L}, \bar{s}) - \bar{\theta}^a,$$
$$\bar{\ell}^a, \bar{s}) - \bar{\lambda}^a \|,$$

$$\Delta_5 = \max_{\bar{s} \in \mathcal{S}, \bar{z} \in \mathcal{F}_{\bar{s}}, a=0,...,N} \| u_\ell(\bar{\theta}^a_- f_K(\bar{K}, \bar{L}, \bar{s}) + (\mathbf{e}^a(\bar{s}) - \bar{\ell}^a) f_L(\bar{K}, \bar{L}, \bar{s}) - \bar{\theta}^a,$$
$$\bar{\ell}^a, \bar{s}) - \bar{\lambda}^a f_L(\bar{K}, \bar{L}, \bar{s}) \|.$$

Given that necessary perturbations in λ are bounded by $\delta_1 + \delta_2^a$, maximal necessary perturbations in Bernoulli utilities are then bounded by $\delta_1 + \delta_2^a + \Delta_4$ for $\bar{w}_1^u$ and $\Delta_5 + (\delta_1 + \delta_2^a) P_\ell^{\max}$ for $\bar{w}_2^u$, where $P_\ell^{\max} = \max_{\bar{s} \in \mathcal{S}, \bar{z} \in \mathcal{F}_{\bar{s}}} f_L(\bar{K}, \bar{L}, \bar{s})$.

Can we find a δ_1 such that for all $\sigma \in \Sigma$ and the exact equilibrium values $\hat{\lambda}$, $\|\hat{\lambda}(\sigma) - \bar{\lambda}\| < \delta_1$? To bound the perturbations of Lagrange multipliers over time, the crucial insight is that we can set $\hat{\lambda}^0 = \bar{\lambda}^0$ and then only must keep track of the propagation of perturbations over an agents' lifetime. For this, define

$$M = \min_{\bar{s}, \bar{z} \in \mathcal{F}_{\bar{s}}} \beta \sum_{s \in \mathcal{S}} \Pi(s, \bar{s}) \rho_{p_k}(\bar{\theta}, s) \quad \text{and} \quad \Delta_6 = \delta_2 \sum_{a=1}^{N} \frac{\delta_2^a}{M^{N-a+1}}.$$

Cumulative perturbations in Lagrange multipliers over an agent's life-span are then bounded by Δ_6. As a result, the cumulative perturbations in Bernoulli utilities are bounded by $\Delta_6 + \Delta_4$ for $\bar{w}_1^u$ and $\Delta_5 + \Delta_6 P_\ell^{\max}$ for $\bar{w}_2^u$. We have now established the maximal necessary perturbations of the ε-equilibrium process in order to obtain an exact equilibrium for a close-by economy.

Error Bound 1. Consider an ε-equilibrium process $(z^\varepsilon(\sigma))$ for the OLG production economy with $z^\varepsilon(\sigma) = (\theta_-^\varepsilon(\sigma), \theta^\varepsilon(\sigma), c^\varepsilon(\sigma), \ell^\varepsilon(\sigma), \lambda^\varepsilon(\sigma), K^\varepsilon(\sigma),$ $L^\varepsilon(\sigma), p_k^\varepsilon(\sigma), p_\ell^\varepsilon(\sigma))$. There exists an economy in equilibrium $(\mathcal{E}((w(\sigma))_{\sigma \in \Sigma}),$ $(\hat{z}(\sigma))_{\sigma \in \Sigma})$ with backward perturbations

$$\|w_1^u(\sigma))_{\sigma \in \Sigma}\| \leq \Delta_6 + \Delta_4 \quad \text{and} \quad \|w_2^u(\sigma))_{\sigma \in \Sigma}\| \leq \Delta_5 + \Delta_6 P_\ell^{\max}$$

and forward perturbations

$$\|(p^\varepsilon(\sigma) - \hat{p}(\sigma))_{\sigma \in \Sigma}\| \le \max(\Delta_1, \Delta_2), \quad \|(c^\varepsilon(\sigma) - \hat{c}(\sigma))_{\sigma \in \Sigma}\| \le \Delta_3,$$
$$\|(\lambda^\varepsilon(\sigma) - \hat{\lambda}(\sigma))_{\sigma \in \Sigma}\| \le \Delta_6.$$

The remaining equilibrium variables have the (unperturbed) values from the ε-equilibrium process, so $\hat{\theta}(\sigma) = \theta^\varepsilon(\sigma)$, $\hat{\ell}(\sigma) = \ell^\varepsilon(\sigma)$, $K(\sigma) = K^\varepsilon(\sigma)$, and $\hat{L}(\sigma) = L^\varepsilon(\sigma)$.

In the statement we report all errors as absolute as opposed to relative errors. In many economic applications it is more meaningful to report relative errors, but the same analysis applies except that all errors have to be taken to be relative errors. Furthermore, Lemma 1 transforms the bounds on utility perturbations into corresponding perturbations of preferences.

If the economic application commands that preferences have to be held fixed, that is, $w_1^{ua}(s) = w_2^{ua}(s) = 0$ for all a, s, then we must perturb both individual endowments and production functions to achieve an approximation to an exact equilibrium. We start with the ε-equilibrium process $(z^\varepsilon(\sigma))$ and ask under which conditions this process approximates an equilibrium process for an economy where endowments and technology are perturbed but utility functions are not. We use the same strategy as above to bound necessary perturbations of λ over time and obtain a value for Δ_6. However, now, for $(h_{\mathrm{olg}}4)$ and $(h_{\mathrm{olg}}5)$ to hold with equality, since $w^u = 0$, consumption and leisure choices have to be different. Bounds δ^c and δ^ℓ on the perturbations in (c, ℓ) can be obtained through a direct calculation:

$$u_c(\bar{c}^a + \delta^c, \bar{\ell}^a + \delta^\ell, \bar{s}) = u_c(\bar{c}^a, \bar{\ell}^a, \bar{s}) + \Delta_6,$$
$$u_\ell(\bar{c}^a + \delta^c, \bar{\ell}^a + \delta^\ell, \bar{s}) = u_\ell(\bar{c}^a, \bar{\ell}^a, \bar{s}) + \Delta_6 P_\ell^{\max}.$$

This directly implies the necessary w^{ea} to make $(h_{\mathrm{olg}}3)$ hold with equality. With different ℓ^a and $w^{ea} \ne 0$, $(h_{\mathrm{olg}}9)$ will no longer hold with equality, and one must perturb $\bar{L}$. Finally, to ensure that $(h_{\mathrm{olg}}6)$ and $(h_{\mathrm{olg}}7)$ hold with equality, given the original prices (which do not need to be perturbed), one must perturb production functions via w^f. Note that we do not perturb endowments of the consumption and capital good, since they are zero.

5.3 Parametric Examples

We illustrate the result of Error Bound 1 with an example. There are $S = 2$ shocks, which are i.i.d. with $\pi_s = 0.5$ for $s = 1, 2$. Suppose the risky spot production function is Cobb–Douglas, $f(K, L, s) = \eta(s)K^\alpha L^{1-\alpha} + (1 - \delta)K$, with $\alpha = 0.36$ and $\delta = 0.7$ for shocks $\eta = (\eta(1), \eta(2)) = (0.85, 1.15)$. Agents live for six periods, $a = 0, 1, ..., 5$, and only derive utility from the consumption good. An agent born at shock s^t has utility function

$$U^{s^t} = E_{s^t} \sum_{a=1}^{N} \beta^{a-1} \frac{(c(s^{t+a-1}))^{1-\gamma}}{1-\gamma}$$

Table 1. Errors

(e^4, e^5)	$\#\mathcal{F}$	ε	M	Δ_6	$P_\ell^{\max}$	$\Delta c(\varepsilon)$	$d^H(P, P')$
(1,1)	4,193,610	9.97 (−4)	1.065	2.94 (−3)	0.407	1.25 (−4)	2.86 (−3)
(0.1,0.1)	1,855,596	3.75 (−3)	0.828	1.66 (−2)	0.493	2.55 (−4)	3.36 (−3)

with a coefficient of relative risk aversion $\gamma = 1.5$ and discount factor $\beta = 0.8$. Individual labor endowments are deterministic, $(e^0, e^1, ..., e^5) = (1.2, 1.3, 1.4, 1.4, e^4, e^5)$. We consider two different specifications of the endowments for $a = 4, 5$, namely $(e^4, e^5) = (1, 1)$ and $(e^4, e^5) = (0.1, 0.1)$.

Table 1 reports the number of elements in the ε-equilibrium set $\mathcal{F}$ and the maximal errors in equilibrium equations, ε, as well as the quantities M, Δ_6, and $P_\ell^{\max}$ that play key roles in the calculations of the bounds on backward and forward errors. We also report $\Delta c(\varepsilon)$ which denotes the maximal consumption-equivalent error in intertemporal Euler equations (see Judd [Jud98]) and the bound on $d^H(P, P')$ from Lemma 1, the distance between original preferences and perturbed preferences.

Note that the necessary perturbations in fundamentals are quite small throughout and are not much larger than the equation error ε. Although for the second specification, perturbations in utilities are quite large (greater than 10^{-2}), translated to consumption-equivalent perturbations they become small.

6 The Lucas Model with Several Agents

As a second application we consider the model of Duffie et al. ([DGMM94, Section 3]. This model is a version of the Lucas [Luc78] asset pricing model with finitely many heterogeneous agents. There are I infinitely lived agents, $i \in \mathcal{I}$, and a single commodity in a pure exchange economy. Each agent $i \in \mathcal{I}$ has endowments $e^i(\sigma) > 0$ at all nodes $\sigma \in \Sigma$, which are time-invariant functions of the shock alone, i.e., there exist functions $\mathbf{e}^i : \mathcal{S} \to \mathbb{R}_+$ such that $e^i(s^t) = \mathbf{e}^i(s_t)$. Agent i has von Neumann–Morgenstern utility over infinite consumption streams $U^i(c) = E_0 \sum_{t=0}^{\infty} \beta^t u_i(c_t)$ for a differentiable, strictly increasing, and concave Bernoulli function u_i which satisfies an Inada condition.

There are J infinitely lived assets in unit net supply. Each asset $j \in \mathcal{J}$ pays shock dependent dividends $d_j(s)$. We denote its price at node s^t by $q_j(s^t)$. Agents trade these assets but are restricted to hold nonnegative amounts of each asset. We denote portfolios by $\theta^i \geq 0$. At the root node s_0, agents hold initial shares $\theta^i(s_{-1})$, which sum to 1.

A competitive equilibrium is a collection $((\hat{c}^i(\sigma), \hat{\theta}^i(\sigma))_{i \in \mathcal{I}}, \hat{q}(\sigma))_{\sigma \in \Sigma}$ such that markets clear and such that agents optimize, i.e.,

$$(\hat{c}^i, \hat{\theta}^i) \in \underset{(c^i, \theta^i) \geq 0}{\arg\max} \, U^i(c^i) \quad \text{s.t.} \quad \forall s^t \in \Sigma$$

$$c^i(s^t) = \mathbf{e}^i(s_t) + \theta^i(s^{t-1})(\hat{q}(s^t) + d(s_t)) - \theta^i(s^t)\hat{q}(s^t).$$

6.1 The Equilibrium Equations

We define the current endogenous variables to consist of beginning-of-period portfolio holdings, $\theta_- = (\theta_-^1, ..., \theta_-^I)$, new portfolio choices, 0, asset prices, q, individuals' consumptions, $c = (c^1, ..., c^I)$, and individuals' marginal utilities, $m^i = u_i'(c^i)$, i.e., $z = (\theta_-, \theta, q, c, m)$. We again build normalizations into the state space so that $\theta^i, c^i \geq 0$ for all $i \in \mathcal{I}$ and that $\sum_{i \in \mathcal{I}} \theta_-^i = 1$.

For the error analysis, we perturb the per-period utility functions, ui, as well as individual endowments. (The error analysis is simplified by considering both perturbations, but we show below that, in general, perturbations only in endowments suffice.) We take perturbations to be $2I$ vectors, $w = (w^u, w^e) = (w^{u1}, ..., w^{uI}, w^{e1}, ..., w^{eI}) \in \mathbb{R}^I \times \mathbb{R}^I$. The equilibrium equations are then $h(\bar{s}, \bar{z}, \bar{w}, \kappa, z(1), ..., z(S)) = 0$ with

$$h = \begin{cases} -\bar{q}\bar{m}^i + \beta \sum_{s \in \mathcal{S}} \Pi(s|\bar{s})(q(s) + d(s))m^i(s) + \kappa^i & (i \in \mathcal{I}) & (h^1), \\ \kappa_j^i \bar{\theta}_j^i & (i \in \mathcal{I}, j \in \mathcal{J}) & (h^2), \\ \bar{m}^i - (u_i'(\bar{c}^i) + \bar{w}^{ui}) & (i \in \mathcal{I}) & (h^3), \\ \bar{c}^i - \bar{\theta}_-^i(\bar{q} + d(\bar{s})) + \bar{\theta}^i \cdot \bar{q} - (\mathbf{e}^i(\bar{s}) + \bar{w}^{ei}) & (i \in \mathcal{I}) & (h^4), \\ \theta_-^i(s) - \bar{\theta}^i & (i \in \mathcal{I}) & (h^5), \\ \sum_{i \in \mathcal{I}} \bar{\theta}_j^i - 1 & (i \in \mathcal{I}, j \in \mathcal{J}) & (h^6). \end{cases}$$

We denote equation $h^i(\cdot) = 0$ by $(h_{\inf}i)$ for all $i = 1, ..., 6$. Duffie et al. [DGMM94] provide conditions on u^i which ensure that these equations are necessary and sufficient for an equilibrium. The natural endogenous state for this economy consists of beginning-of-period portfolios, see, e.g., Heaton and Lucas [HL96], although a recursive equilibrium for this state space cannot be shown to exist (see Duffie et al. [DGMM94]).

For reasons that will become clear in the error analysis below, we need a slightly extended state space. For a small $\eta > 0$, define the state space

$$\Psi^\eta = \left\{ \theta \in \mathbb{R}^{IJ} : \sum_{i \in \mathcal{I}} \theta_j^i = 1, \; \theta_j^i \geq -\eta \text{ for all } i \in \mathcal{I}, j \in \mathcal{J} \right\}$$

and let $\rho = ((\rho_{\theta^i})_{i \in \mathcal{I}}, \rho_q, (\rho_{c^i}, \rho_{m^i})_{i \in \mathcal{I}}$ denote the policy function for a recursive ε-equilibrium. For sufficiently small η, recursive ε-equilibria exist (see Kubler and Schmedders [KS03]). We denote the equilibrium set by $\mathcal{F}$.

6.2 Error Analysis

We must perturb marginal utility in order to set $h^1 = 0$ for a given ε-equilibrium process. However, because agents are infinitely lived, the main problem in the error analysis is that the necessary perturbations to correct for errors in $(h_{inf}1)$ along the event tree may propagate without bounds. Figure 3 illustrates this problem for the economy in Example 2. The asset holdings of the ε-equilibrium process diverge from the exact equilibrium. Setting the errors in the equilibrium equations to zero then requires increasingly larger deviations of marginal utilities. The backward substitution approach of the previous section does not yield small bounds on the necessary perturbations, and we are no longer able to show that the ε-equilibrium is path-approximating. Instead we show that the ε-equilibrium weakly approximates an economy in equilibrium. Intuitively, a weak approximation only demands that there is an exact equilibrium that can be approximately generated by the computed policy function of the recursive ε-equilibrium. Contrary to path-approximation, this concept allows for perturbations in the state variable. Figure 3 displays the basic idea for such an error analysis. At a perturbed value of the state variable, the approximate policy function yields the exact equilibrium variables for the current period. (The exactness in the figure is an idealization: in general, the perturbation of state variables will not yield an exact solution to the equilibrium equations, and so some additional forward and backward perturbations are necessary.) The advantage of the evaluation of the policy function at a close-by value is that the magnitude of all perturbations remains small and tight bounds for a weak approximation can be established. We now formalize the depicted intuition for the Lucas model.

To simplify the analysis, we assume that for the ε-equilibrium, $\mathcal{F}$, $(h_{inf}3)$–$(h_{inf}6)$ actually hold with equality, so for all values in $\mathcal{F}$ market clearing and budget feasibility are built in. As in the OLG model, actual errors in these equations will be negligible.

To construct a bound δ that ensures a weakly approximating ε-equilibrium, we show the stronger result that there exists an exact equilibrium $(\hat{z}(\sigma))_{\sigma \in \Sigma}$, such that for all $s^t \in \Sigma$, there exist some $z \in \mathcal{F}_{s_t}$ with $\|z - \hat{z}(s^t)\| < \delta$, which differs from $\hat{z}$ only in m and θ_-. No other variables are perturbed, and so $\hat{\theta}(s^t) = \theta$, $\hat{q}(s^t) = q$, $\hat{c}(s^t) = c$. For the budget constraint (h_{inf}) to hold for the exact equilibrium value $\hat{z}(s^t)$, one must perturb endowments, that is, allow we to be nonzero. This perturbation is necessitated by the fact that for the ε-equilibrium values the budget constraint is assumed to hold exactly, and so perturbations in θ_- introduce an error in this equation. For these errors to be less than δ, we need to require that $\|(\hat{\theta}^i_-(s^t) - \theta_-)(q + d(s_t))\| < \delta$. As in Example 2, the strategy is to find perturbations in the endogenous state that ensure the value of the approximate policy function almost matches the exact equilibrium value.

A sufficient condition for the existence of a close-by economy in equilibrium is then that for every $\bar{z}^P$ that is identical to some $\bar{z} \in \mathcal{F}_{\bar{s}}$ in all coordinates with the exception of $\bar{m}$ and $\bar{\theta}_-$ where it has the property that $\|\bar{m}^P - \bar{m}\| < \delta$ and $\|(\bar{\theta}^P_- - \bar{\theta}_-)(\bar{q} + d(\bar{s}))\| < \delta$, we can find consistent values of next period's endogenous variables, which are identical to elements of $z(s) \in \mathcal{F}_s$, $s \in \mathcal{S}$, except that again the m's and θ_-'s satisfy the above condition. More formally, suppose that for all $\bar{s}$, $\bar{z} \in \mathcal{F}_{\bar{s}}$ and all $\bar{z}^P$ with $\|\bar{z}^P - \bar{z}\| < \delta$ and $\bar{q}^P = \bar{q}$, $\bar{\theta}^P = \bar{\theta}$, $\bar{c}^P = \bar{c}$, $\|(\bar{q} + d(\bar{s}))(\bar{\theta}_- - \bar{\theta}_-^P)\| < \delta$ there exist $\tilde{z}(1), ..., \tilde{z}(S)$ such that:

(I1) For all $s \in \mathcal{S}$ it holds that $\|z(s) - \tilde{z}(s)\| < \delta$ for some $z(s) \in \mathcal{F}_s$ with

$$\tilde{q}(s) = q(s), \quad \bar{\theta}(s) = \theta(s), \quad \bar{c}(s) = c(s),$$
$$\|(q(s) + d(s))(\tilde{\theta}_-(s) - \theta_-(s))\| < \delta$$

(I2) For some $\kappa \in \mathcal{K}$ and w with $\|w\| < \delta$, $h(\bar{s}, \bar{z}^P, w, \kappa, \tilde{z}(1), ..., \tilde{z}(S)) = 0$.

Then there exists an economy in equilibrium $(\mathcal{E}((w(\sigma))_{\sigma \in \Sigma}), (\hat{z}(\sigma))_{\sigma \in \Sigma})$, with $\sup_{\sigma \in \Sigma} \|w(\sigma)\| < \delta$ such that for all $s^t \in \Sigma$ and all $\hat{z}(s^t)$ it holds that $\|z - \hat{z}(s^t)\| < \delta$ for some $z \in \mathcal{F}_{s_t}$.

The rather technical conditions (I1) and (I2) nicely correspond to our intuition of a sensible error analysis for the infinite-horizon model. Given a previously perturbed point $\bar{z}^P$, we can find a perturbation $\tilde{z}$ of some point z in the equilibrium set $\mathcal{F}$ that sets the equilibrium equations at $(\bar{s}, \bar{z}^P)$ equal to zero for a slightly perturbed ($\|w\| < \delta$) economy. An iterated application of this process then leads to an exact equilibrium process $\hat{z}$ for a close-by economy.

To verify the conditions for a given ε-equilibrium, we need some additional notation. For an under determined system $Ax = b$ with a matrix A that has linearly independent rows, denote by $A^+ = A^\top (AA^\top)^{-1}$ the pseudo inverse of A. The unique solution of the system that minimizes the Euclidean norm $\|x\|_2$ is then given by $x_{LS} = A^+ b$. We use the Euclidean norm here since it is well understood how to compute $A^+ b$ accurately. The approach immediately yields an upper bound on the sup-norm of the error since $\|x\| \leq \|x\|_2$ for $x \in \mathbb{R}^n$. Using the fact that we consider a recursive ε-equilibrium, we can write current endogenous variables as functions of θ_- alone and define for any current shock $\bar{s}$ and any $\theta_-(1), ..., \theta_-(S) \in \Psi^n$ a $J \times S$ payoff matrix by

$$M(\bar{s}, \vec{\theta}_-) = M(\bar{s}, \theta_-(1), ..., \theta_-(S))$$
$$= \left(\beta \Pi(s|\bar{s})(\rho_{q_j}(s, \theta_-(s)) + d_j(s)) \right)_{js}.$$

For any $\bar{m}^P$, $\bar{s}$, $\bar{z} \in \mathcal{F}_{\bar{s}}$, $\kappa \in \mathbb{R}_+^J$, and $\vec{\theta}_-$, define S-dimensional vectors for each agent i:

$$E^i(\bar{s}, \bar{m}^P, \bar{q}, \kappa, \vec{\theta}_-)$$
$$= (M(\bar{s}, \vec{\theta}_-))^+ \left[\bar{q}\bar{m}^{P_i} - M(\bar{s}, \vec{\theta}_-) \begin{pmatrix} u_i'(\rho_{c^i}(1, \theta_-(1))) \\ \vdots \\ u_i'(\rho_{c^i}(S, \theta_-(S))) \end{pmatrix} - \kappa \right].$$

These are the necessary perturbations in $m^i(1), ..., m^i(S)$ for $(h_{\inf}1)$ to hold with equality if the next period's states are $\vec{\theta}$ and this period's endogenous variables are identical to $\bar{z}$, except that this period's marginal utilities might

Table 2. Errors

ε	$w^u_{\max}$	$w^e_{\max}$	$\bar{w}^e_{\max}$
2.04 (−3)	3.53 (−3)	1.19 (−3)	4.2 (−3)

differ from $\bar{m}$ and be given by $\bar{m}^P$. The formula appears complicated only because we consider the general case of J assets and S states. In the example below it simplifies considerably. Lemma 2 gives sufficient conditions to ensure that (I1) and (I2) hold.

Lemma 2. *Given $\delta > 0$, suppose that for all $\bar{s} \in \mathcal{S}$, $\bar{z} \in \mathcal{F}_{\bar{s}}$, and all $(n_1, ..., n_I) \in \{-1, 1\}^I$, there exists $\vec{\theta}_- \in (\Psi^\eta)^S$ with $\max_{i \in \mathcal{I}, s \in \mathcal{S}} \|(\theta^i_-(s) - \bar{\theta}^i)(\rho_q(s, \theta_-(s)) + d(s))\| \le \delta$ such that for all $i \in \mathcal{I}$ there exists a $\kappa \ge 0$ with $\kappa \bar{\theta}^i = 0$, which ensures that for all $s \in \mathcal{S}$,*

$$\|E^i_s(\bar{s}, \bar{m}, \bar{q}, \kappa, \vec{\theta}_-)\| < \min(\delta, \delta u'_i(\rho_{c^i}(s, \theta_-(s)))) \quad and$$
$$\operatorname{sgn}(E^i_s(\bar{s}, \bar{m}, \bar{q}, \kappa, \vec{\theta}_-)) = n_i.$$

Then there exists an economy in equilibrium $(\mathcal{E}((w(\sigma))_{\sigma \in \Sigma}), (\hat{z}(\sigma))_{\sigma \in \Sigma})$, with $\sup_{\sigma \in \Sigma} \|w(\sigma)\| < \delta$ such that for all $s^t \in \Sigma$ and all $\hat{z}(s^t)$ it holds that $\|z - \hat{z}(s^t)\| < \delta$ for some $z \in \mathcal{F}_{s_t}$.

For all $\vec{\theta}_- \in \Psi^\eta$ and all $\bar{s} \in \mathcal{S}$ with associated $\bar{q}$, $\bar{m}$, $\bar{\theta}$, one can perform a grid search to verify the conditions of the lemma and to determine a bound on backward errors in endowments, $w^e_{\max}$ as well as errors in utilities, $w^u_{\max}$. We illustrate below some problems that arise in practice.

If one wants to perturb endowments only, one must translate the perturbations in marginal utility into perturbations in consumption values and perturb individual endowments to satisfy the budget constraints. A crude bound can be obtained by computing

$$\max_{s \in \mathcal{S}, i \in \mathcal{I}, \theta \in \Theta} \{u'^{-1}_i(u'_i(\rho_{c^i}(s, \theta)) + w^u_{\max})$$
$$- \rho_{c^i}(s, \theta), u'^{-1}_i(u'_i(\rho_{c^i}(s, \theta)) - w^u_{\max}) - \rho_{c^i}(s, \theta)\}.$$

The total necessary perturbation is given by the sum of this expression and $w^e_{\max}$. In Table 2 we denote this by $\bar{w}^e_{\max}$.

6.3 Parametric Example

The following small example illustrates the analysis above. There are $S = 2$ shocks, which are i.i.d. and equiprobable. There are two agents with CRRA utility functions that have identical coefficient of risk aversion of $\gamma = 1.5$ who discount the future with $\beta = 0.95$. Individual endowments are $e^1 = (1.5, 3.5)$ and $e^2 = (3.5, 1.5)$. There is a single tree with dividends $d(s) = 1$ for $s = 1, 2$.

Since there are only two agents, the endogenous state space for the recursive ε-equilibrium simply consists of the interval $[0, 1]$. For our error analysis it is crucial, however, that we can perturb $\theta_-(s)$ even if this period's choice is $\bar{\theta} = 0$. For this, we extend the state space to $[-0.01, 1.01]$. The bounds should be chosen to guarantee that nonnegative consumption is still feasible at all points in the state space. Standard algorithms (see, e.g., Kubler and Schmedders [KS03]) can be used to compute a recursive ε-equilibrium even for this extended state space. At $\bar{\theta}_- < 0$, the short-sale constraint forces the new choice $\rho_\theta(\bar{s}, \bar{\theta})$ to be nonnegative, no matter what the current shock, and our error analysis as outlined above goes through. We obtain the errors reported in Table 2.

7 Conclusion

An error analysis should ideally relate the computed solution to an exact equilibrium of the underlying economy. However, unfortunately this is generally impossible. Instead, we argue that it is often economically interesting to relate the computed solution to an exact solution of a close-by economy and to perform a backward error analysis.

For stochastic infinite-horizon models we define economies to be close by if preferences and endowments are close by in the sup-norm. With this definition, we show how to construct ε-equilibrium processes from computed candidate solutions that approximate an exact equilibrium of a close-by economy. When agents are finitely lived, this construction is straightforward, as the process is generated by the transition function of the recursive approximate equilibrium. When agents are infinitely lived, the construction is more elaborate, since one cannot guarantee sample-path stability of the equilibrium transition. In practice one needs to be content with a notion of "weak approximation."

Acknowledgment

We thank seminar participants at various universities and conferences, and especially Don Brown, Bernard Dumas, John Geanakoplos, Peter Hammond, Martin Hellwig, Ken Judd, Mordecai Kurz, George Mailath, Alvaro Sandroni, Manuel Santos, and Tony Smith, for helpful discussions. We are grateful to the associate editor and three anonymous referees for useful comments.

Kubler, F., Schmedders, K.: Approximate versus exact equilibria in dynamic economies. Econometrica **73**, 1205–1235 (2005). Reprinted by permission of the Econometric Society.

Appendix

Proof of Lemma 1. Define $z \in \ell_\infty$ by $z_\ell(s^t) = \delta_\ell(s^t) x_\ell(s^t)$ for all ℓ, s^t. For $y \in \ell_\infty$, define $t(y) = \sum_{t=0}^\infty \beta^t \sum_{s^t} \pi(s^t)(D_x u(x(s^t), s_t) + w(s^t)) \cdot y(s^t)$. The closed half-space $\{y : t(y) > t(x)\}$ does not contain any point from the interior of $\mathcal{B}$. Postlewaite and Schmeidler ([PS81], p. 109) construct an indifference curve that is identical to the indifference curve of U passing through x in this half-space and that is identical to the boundary of the half-space in the region where the indifference curve of U is outside of the half-space. By their construction, to prove the theorem, it suffices to derive an upper bound on

$$\varepsilon = d^H (\{y : U(y) \geq U(x)\},\ \{y : t(y) \geq t(x) \wedge U(y) \geq U(x)\}).$$

Observe that the maximum will be obtained at some $\bar{y}$ satisfying $U(\bar{y}) = U(x)$ and $D_x U(\bar{y})$ collinear to $D_x U(x) + w$. By the definition of d^H, we must have $\|\bar{y} - x\| \geq \varepsilon$. By homotheticity, there exists a scalar $\alpha > 0$ such that $\bar{y} = \alpha z$. Since monotonicity of preferences implies that $\bar{y}_\ell(s^t) \geq x_\ell(s^t)$ for some ℓ, s^t, $\|\bar{y} - x\|$ must be bounded by $\|\bar{z} - x\|$, where $\bar{z} = z / \sup_\sigma \|\delta(\sigma)\|$. This is true because $\bar{z} = \bar{\alpha} z$ for some $\bar{\alpha}$ and $\bar{z} \leq \bar{y}$. By definition of z it follows that $\varepsilon \leq \sup_{s^t, \ell} x_\ell(s^t)(1 - \delta_\ell(s^t)/ \sup_\sigma \|\delta(\sigma)\|)$. □

Proof of Lemma 2. To prove the lemma, it is useful to consider relative errors as opposed to absolute errors and to define $R^i(\bar{s}, \bar{m}, \bar{q}, \kappa, \vec{\theta}_-)$ by $R^i_s = E^i_s/u'_i(\rho_{c^i}(s, \theta_-(s)))$ for all $s \in \mathcal{S}$. With this definition,

$$\bar{q}\bar{m}^{P_i} - M(\bar{s}, \vec{\theta}_-) \begin{pmatrix} (1 + R^i_1)u'_i(\rho_{c^i}(1, \theta_-(1))) \\ \vdots \\ (1 + R^i_S)u'_i(\rho_{c^i}(S, \theta_-(S))) \end{pmatrix} - \kappa = 0.$$

It follows that for any scalar $\gamma > -1$, there is some $\kappa' \geq 0$, $\kappa'\bar{\theta}^i = 0$ such that for any $s \in \mathcal{S}$,

$$\|R^i_s(\bar{s}, (1+\gamma)\bar{m}^P, \bar{q}, \kappa', \vec{\theta})\| \leq \|(1+\gamma)(1 + R^i_s(\bar{s}, \bar{m}^P, \bar{q}, \kappa, \vec{\theta}_-)) - 1\|. \quad (1)$$

Given $\bar{s}, \bar{z} \in \mathcal{F}_{\bar{s}}$ and a perturbed $\bar{z}^P$ with $\bar{q}^P = \bar{q}$, $\|\bar{m}^P - \bar{m}\| < \delta$, we can write $\bar{m}^{iP} = (1 + \gamma_i)\bar{m}^i$ for some $\gamma = (\gamma_1, ..., \gamma_I)$ with $\|\gamma\| \leq \delta$. Given $(\text{sgn}(\gamma_1), ..., \text{sgn}(\gamma_I)) \in \{-1, 1\}^I$, the conditions of the lemma require that

$$\|R^i_s(\bar{s}, \bar{m}, \bar{q}, \kappa, \vec{\theta}_-)\| \leq \min\left(\delta, \frac{\delta}{u'_i(\rho_{c^i}(s, \theta_-(s)))}\right) \quad \text{and}$$

$$\text{sgn}(R^i_s(\bar{s}, \bar{m}, \bar{q}, \kappa, \vec{\theta}_-)) = -\text{sgn}(\gamma_i).$$

With (1), because of the sign condition, we obtain $\|R^i_s(\bar{s}, \bar{m}(1+\gamma_i), \bar{q}, \kappa, \vec{\theta}_-)\| \leq \min(\delta, \delta/u'_i(\rho_{c^i}(s, \theta_-(s))))$ for all $s \in \mathcal{S}$. This shows that errors do not propagate and therefore proves the lemma. □

Tame Topology and *O*-Minimal Structures

Charles Steinhorn

Vassar College, Poughkeepsie, NY 12604 `steinhorn@vassar.edu`

I first want to thank Professor Brown for his kind introduction. My goal is to give everyone a sense of the subject of o-minimality. Ten hours is a lot of time to lecture, but the literature on this topic has grown vast, so I will ignore a lot of things in order to get at issues relevant for this audience. I confess that this is the first time I have spoken to economists.

I want to start with an outline of the talks. The topic is tame topology and o-minimal structures, but today I'm not going to tell you what either one really is. Today I'm going to give you a sense of what's important so that you can understand things better in my subsequent lectures.

Outline of Lectures

1. An introduction to definability and quantifier elimination
2. The semialgebraic case
3. *O*-minimality and some basic properties
4. Examples and some further properties
5. VC dimension and applications

1 An Introduction to Definability

For which x do we have
$$\exists y \begin{bmatrix} ax + by > e_1 \\ cx + dy \le e_2 \end{bmatrix} \qquad (*)$$
where $a, b, c, d, e_1, e_2 \in \mathbb{R}$? We want to know
$$\{x \in \mathbb{R} \mid \text{statement (1) holds for } x\}.$$

We also can replace the constants a, b, c, d, e_1, e_2 by variables u, v, w, z, s_1, s_2 and ask for which $u, v, w, z, s_1, s_2 \in \mathbb{R}$ do we have

$$\exists x \exists y \begin{bmatrix} ux + vy > s_1 \\ wx + zy \le s_2 \end{bmatrix}. \qquad (**)$$

That is, we want to know

$$\{(u, v, w, z, s_1, s_2) \in \mathbb{R}^6 \,|\, (u, v, w, z, s_1, s_2) \text{ satisfies (1)}.$$

These are simple examples of *definable sets*. I want to give a couple more examples before I start to get precise.

Examples

1. $S = \{x \in \mathbb{R} \,|\, \sin \pi x = 0\}$. Observe $S = \mathbb{Z}$.
2. Let $L = \{(m, n) \in \mathbb{R}^2 \,|\, m, n \in \mathbb{Z}\}$ be the integer lattice in $\mathbb{R}^2$. View L as a two-place relation on pairs of real numbers:

$$L(m, n) \iff m, n \in \mathbb{Z}.$$

Define

$$C = \left\{ (x, y) \in \mathbb{R}^2 \,|\, \exists u \exists v [L(u, v) \wedge (x - u)^2 + (y - v)^2 \le 1/16] \right\}.$$

The set C is the set of all closed discs of radius $1/4$ whose centers are points in L.

You can see that as you introduce quantifiers, you have to think a bit about what you are defining.

1.1 Structures on the Real Numbers $\mathbb{R}$

Throughout these lectures, I am going to limit my discussion to structures on the real line, for concreteness. Fix a family of

1. $\mathcal{F}$ of *basic functions* $f : \mathbb{R}^k \to \mathbb{R}$, for $k = 1, 2, 3, \dots$.
2. $\mathcal{R}$ of *basic relations* (subsets) $R \subseteq \mathbb{R}^k$, for $k = 1, 2, 3, \dots$, such that the "less than" relation $< \subset \mathbb{R}^2$ (and equality) always is included in $\mathcal{R}$.

Refer to $\mathcal{F}$ and $\mathcal{R}$ together as a language $\mathcal{L}$. The real numbers $\mathbb{R}$ together with the functions and relations included in $\mathcal{L}$ is called an $\mathcal{L}$-*structure* that in general we denote by $\mathfrak{R}_\mathcal{L}$. Special structures will have special names.

Examples

1. $\mathcal{F} = \{+, \cdot, -\}$ where $- : \mathbb{R} \to \mathbb{R}$ is given by $x \longmapsto -x$ and $\mathcal{R} = \{<\}$. This is the language $\mathcal{L}_{\text{alg}}$ of the ordered field of real numbers, the structure denoted by $\mathbb{R}_{\text{alg}}$.

2. $\mathcal{F}$ contains $+$ and, for each $r \in \mathbb{R}$, the scalar multiplication function $\mu_r : \mathbb{R} \to \mathbb{R}$ given by $x \longmapsto rx$, and $\mathcal{R} = \{<\}$. This is the ordered real vector space language $\mathcal{L}_{\mathrm{lin}}$ for the real numbers, whose corresponding structure is denoted by $\mathbb{R}_{\mathrm{lin}}$.
3. $\mathcal{F} = \{+, \cdot, -, \exp\}$ where $\exp : \mathbb{R} \to \mathbb{R}$ is the exponential function $x \longmapsto e^x$ and $\mathcal{R} = \{<\}$. This is the language $\mathcal{L}_{\exp}$ of the ordered exponential field of real numbers; the structure is denoted by $\mathbb{R}_{\exp}$.

1.2 Terms

Construct an expanded class of functions by repeated composition starting from the functions in $\mathcal{F}$. These are called *terms*.

Examples

1. The $\mathcal{L}_{\mathrm{alg}}$-terms are all integer coefficient polynomial functions $p(x_1, ..., x_k)$ for $k = 1, 2, 3, ...,$
2. The $\mathcal{L}_{\mathrm{lin}}$-terms are all $\mathbb{R}$-linear functions $L(x_1, ..., x_k) = \mu_{r_1} x_1 + \cdots + \mu_{r_k} x_k$.

The obvious kinds of rules apply in these examples, for instance $\mu_{r_1}(x) + \mu_{r_2}(x) = \mu_{r_1 + r_2}(x)$.

1.3 Formulas

If you think of the definable sets I wrote down at the beginning of the lecture, they were defined in terms of equalities and inequalities. The basic formulas are

1. $t_1 = t_2$ for terms t_1, t_2.
2. $R(t_1, ..., t_k)$ where R is a k-place relation and $t_1, ..., t_k$ are terms. Note that $t_1 < t_2$ is a special case.

So if you think of $\mathbb{R}_{\mathrm{alg}}$, we can write down polynomial equalities and inequalities, and for $\mathbb{R}_{\mathrm{lin}}$ we can write down linear equalities and inequalities. Next, from the basic formulas recursively construct

1. by Boolean operations from formulas φ and ψ :

$$\varphi \wedge \psi \text{ read as "}\varphi \text{ and } \psi\text{,"}$$
$$\varphi \vee \psi \text{ read as "}\varphi \text{ or } \psi\text{,"}$$
$$\neg\varphi \quad \text{ read as "not } \varphi\text{."}$$

2. by existential quantification over $\mathbb{R}$ from a formula φ:

$$\exists v \varphi \text{ read as "there exists } v \text{ such that } \varphi\text{."}$$

Caution: to get to the point more quickly I have mixed syntax and semantics. When you look at a formal text on model theory you will see a distinction between symbols in a formal language and their interpretations.

Now what I'm going to do is go back to what I was vague about earlier: $\mathcal{L}$-definable sets.

$\mathcal{L}$-definable sets

For our purposes these are subsets of $\mathbb{R}^k$, for $k = 1, 2, 3, ...$, specified as follows. For each $\mathcal{L}$-formula φ, certain variables are bound to quantifiers and others are not. Call the latter *free variables*, and it is for these that we can substitute real numbers.

For an $\mathcal{L}$-formula φ list its free variables as $x_1, ..., x_k, z_1, ..., z_m$. Choose $c_1, ..., c_m \in \mathbb{R}$ and substitute them for $z_1, ..., z_m$, respectively. Write $\bar{c}$ for $(c_1, ..., c_m)$. The set

$$D_{\varphi, \bar{c}} = \{(x_1, ..., x_k) \in \mathbb{R}^k \mid \varphi(x_1, ..., x_k, \bar{c}) \text{ is true}\} \subseteq \mathbb{R}^k$$

is an $\mathcal{L}$-definable subset of $\mathbb{R}^k$. If $\mathcal{L}$ is clear from context, we usually shall drop the $\mathcal{L}$.

Examples

Let's go back to our earlier examples of definable sets. The set

$$\{x \in \mathbb{R} \mid \text{statement (1) holds for } x\}$$

is $\mathcal{L}_{\text{lin}}$-definable. The set

$$\{(u, v, w, z, s_1, s_2) \in \mathbb{R}^6 \mid (u, v, w, z, s_1, s_2) \text{ satisfies (1)}\}$$

is $\mathcal{L}_{\text{alg}}$-definable. The set

$$\{x \in \mathbb{R} \mid \sin \pi x = 0\}$$

is $(\mathcal{F}, \mathcal{R})$-definable, where $\mathcal{F} = \{\cdot, \sin\}$, $\mathcal{R} = \{<\}$. The set

$$\{(x, y) \in \mathbb{R}^2 \mid \exists u \exists v, [L(u, v) \wedge (x - u)^2 + (y - v)^2 \leq 1/16]\}$$

is $(\mathcal{F}, \mathcal{R})$-definable, where $\mathcal{F} = \{+, \cdot, -\}$, $\mathcal{R} = \{L, <\}$.

Comments

- We call the $\bar{c}$ *parameters*; the division of the free variables in a formula into those which are parameter variables and those that are not can be made arbitrarily.

- A function $F : A \subseteq \mathbb{R}^m \to \mathbb{R}^n$ is said to be $\mathcal{L}$-definable if its graph, as a subset of $\mathbb{R}^m \times \mathbb{R}^n$, is $\mathcal{L}$-definable.
- If a function $F(x_1, ..., x_m)$ is definable, we can treat it as if it is a term for the purposes of constructing more complicated definable functions and formulas.
- Similarly, we can treat a definable subset of $\mathbb{R}^k$ as a basic relation in the language for such purposes.

What I want to do now is consider some further examples of definability, to explore the capabilities of the language we are dealing with.

Some Further Examples

1. Let $f : \mathbb{R} \to \mathbb{R}$ be $\mathcal{L}$-definable, where $\mathcal{L}$ contains $\mathcal{L}_{\text{alg}}$. Then

$$\{x \in \mathbb{R} \mid f \text{ is convex in an interval around } x\}$$

   is $\mathcal{L}$-definable, since "f is convex in an interval around x" can be written as

$$\exists x_1 \exists x_2 (x_1 < x \wedge x < x_2 \wedge \text{``}f \text{ is convex in the interval } (x_1, x_2)\text{''}),$$

   and "f is convex in the interval (x_1, x_2)" can be written as

$$\forall t \left((x_1 < t \wedge t < x_2) \to f(t) < \frac{x_2 - t}{x_2 - x_1} f(x_1) + \frac{t - x_1}{x_2 - x_1} f(x_2) \right).$$

   Note that $\forall t := \neg \exists t \neg$ and $P \to Q := \neg P \vee Q$.

2. Let $f : A \subset \mathbb{R}^k \to \mathbb{R}$ be $\mathcal{L}$-definable, where $\mathcal{L}$ contains $\mathcal{L}_{\text{alg}}$. Then $\{\bar{x} \in A \mid f \text{ is continuous at } \bar{x}\}$ is $\mathcal{L}$-definable, since "f is continuous at x" can be written as

$$\forall \varepsilon (\varepsilon > 0 \to \exists \delta (\delta > 0 \wedge \forall y (|y - x| < \delta) \to |f(y) - f(x)| < \varepsilon)).$$

   Same for differentiability. Note that $|y - x| < \delta := -\delta < y - x \wedge y - x < \delta$.

3. Let $f : \mathbb{R} \to \mathbb{R}$ be $\mathcal{L}$-definable, where $\mathcal{L}$ contains $\mathcal{L}_{\text{alg}}$. If f is differentiable, then f' is $\mathcal{L}$-definable. Same for functions of several variables.

4. Let $f : \mathbb{R} \to \mathbb{R}$ be $\mathcal{L}$-definable, where $\mathcal{L}$ contains $\mathcal{L}_{\text{alg}}$. Then

$$\{x \in \mathbb{R} \mid f \text{ is Lipschitz in an interval around } x\}$$

   is $\mathcal{L}$-definable.

5. Let $A \subset \mathbb{R}^k$ be $\mathcal{L}_{\text{alg}}$-definable. Then there is a formula that expresses that "A is convex." Same for any $\mathcal{L}$ containing $\mathcal{L}_{\text{alg}}$.

6. Let $A \subset \mathbb{R}^k$ be $\mathcal{L}$-definable, where $\mathcal{L}$ is arbitrary. Then the topological closure of A in $\mathbb{R}^k$ is $\mathcal{L}$-definable. Many other notions from point-set topology also are definable.

7. Let $A \subset \mathbb{R}^k$ be $\mathcal{L}$-definable. Then all level sets of A are $\mathcal{L}$-definable.
8. Closed polyhedra in $\mathbb{R}^k$ are $\mathcal{L}_{\text{lin}}$-definable.

Some guidelines for which sets are definable:

- Defining formulas cannot be infinitely long. The formula "$X \subseteq \mathbb{R}$ is finite," which we can write as "X contains 1 element $\lor X$ contains 2 elements $\lor ...$," is not definable.
- Quantification over real numbers only is allowed. Quantifying over the class of all polynomials is not permitted, but quantifying over all polynomials of degree $\leq n$ is allowed, since such polynomials are defined by $n + 1$ coefficients.

1.4 Set-theoretic Definability

The $\mathcal{L}$-definable subsets of $\mathbb{R}^n$ for $n = 1, 2, 3, ...$ is the smallest collection $\mathfrak{D} = \{\mathcal{D}_n \mid n \geq 1\}$ such that

1. Each $D \in \mathcal{D}_n$ is a subset of $\mathbb{R}^n$;
2. $\mathbb{R}^n \in \mathcal{D}_n$;
3. The graph of each $f : \mathbb{R}^n \to \mathbb{R}$ in $\mathcal{F}$ is in $\mathcal{D}_{n+1}$;
4. Each $R \subseteq \mathbb{R}^n$ in $\mathcal{R}$ is in $\mathcal{D}_n$;
5. For all $1 \leq i, j \leq n$, $\{(x_1, ..., x_n) \in \mathbb{R}^n \mid x_i = x_j\} \in \mathcal{D}_n$;
6. Each $\mathcal{D}_n$ is closed under intersection, union, and complement;
7. If $\pi : \mathbb{R}^n \to \mathbb{R}^m$ is a projection map $(x_1, ..., x_n) \longmapsto (x_{i_1}, ..., x_{i_m})$ and $X \in \mathcal{D}_n$ then $\pi(X) \in \mathcal{D}_m$;
8. If π is as above and $Y \in \mathcal{D}_m$ then $\pi^{-1}(Y) \in \mathcal{D}_n$;
9. If $X \in \mathcal{D}_{n+m}$ and $\bar{b} \in \mathbb{R}^m$, then $\{\bar{a} \in \mathbb{R}^n \mid (\bar{a}, \bar{b}) \in X\} \in \mathcal{D}_n$.

1.5 Finer structure of definable sets

A definable set typically has several different definitions. We have the $\mathcal{L}_{\text{alg}}$ definitions of the unit interval $[-1, 1]$:

$$\{x \in \mathbb{R} \mid (-1 < x \land x < 1) \lor x = -1 \lor x = 1\}$$
$$\{x \in \mathbb{R} \mid \exists y \exists z \; x^2 + y^2 + z^2 = 1\}$$
$$\{x \in \mathbb{R} \mid \forall y \exists z \; x^2 + (y - z)^2 = 1\}.$$

These are three different definitions of the same set. We could construct infinitely many. The maxim here is that quantification generates conceptual complexity. One goal of model theory is to attempt to analyze definable sets in a specific context by showing that definable sets can be defined by simple formulas.

1.6 Quantifier elimination for $\mathcal{L}_{\text{lin}}$-definable sets

Theorem. *For every $n = 1, 2, 3, \ldots$, every $\mathcal{L}_{\text{lin}}$-definable subset of $\mathbb{R}^n$ can be defined by a quantifier-free $\mathcal{L}_{\text{lin}}$-formula.*

For convenience, write rx instead of $\mu_r(x)$, where $r \in \mathbb{R}$. The theorem tells us that every $\mathcal{L}_{\text{lin}}$-definable subset of $\mathbb{R}^n$ is a finite Boolean combination (i.e., finitely many intersections, unions, and complements) of sets of the form

$$\{(x_1, \ldots, x_n) \in \mathbb{R}^n \mid a_1 x_1 + \cdots + a_n x_n > b\}$$

where $a_1, \ldots, a_n, b$ are fixed but arbitrary real numbers.

The $\mathcal{L}_{\text{lin}}$-definable sets are called the *semilinear* sets. By routine set theoretic manipulation, semilinear sets can be written as a finite union of the intersection of finitely many sets defined by conditions of the form

$$a_1 x_1 + \cdots + a_n x_n + b = 0,$$
$$c_1 x_1 + \cdots + c_n x_n + d > 0.$$

Thus $\mathcal{L}_{\text{lin}}$-definability reduces (basically) to linear algebra, which we understand well.

$\mathcal{L}_{\text{lin}}$-definable subsets of $\mathbb{R}$

All $\mathcal{L}_{\text{lin}}$-definable subsets of $\mathbb{R}$ are finite Boolean combinations of sets of the form $\{x \in \mathbb{R} \mid ax > b\}$. Geometrically these are the union of finitely many (possibly unbounded) open intervals and points. Consequently, neither $\mathbb{Z}$ nor $\mathbb{Q}$ is $\mathcal{L}_{\text{lin}}$-definable.

Idea of the Proof

Eliminate quantifiers one at a time (proceed inductively).

Example

Eliminate the quantifier for the $\mathcal{L}_{\text{lin}}$-definable set

$$\{x \in \mathbb{R} \mid \exists y, [2x - 3y > 2 \wedge 4x - 2y \leq 0]\}.$$

High school algebraic elimination leads to $\{x \in \mathbb{R} \mid x < 1/2\}$.

Question: What about $\mathcal{L}_{\text{alg}}$-definability? I will discuss this in my next lecture.

2 The Semialgebraic Case

Today I'm going to be discussing the semialgebraic case, that is, the case of $\mathcal{L}_{alg}$. To construct this language, you start with polynomials with real coefficients, and build up through the operations of $+, \cdot, -$. The focus of today's lecture is the Tarski–Seidenberg theorem.

Theorem (Tarski–Seidenberg Theorem [Tar51]). *For every* $n = 1, 2,$
$3, ...,$ *every* $\mathcal{L}_{alg}$*-definable subset of* $\mathbb{R}^n$ *can be defined by a quantifier-free* $\mathcal{L}_{alg}$*-formula.*

Thus every $\mathcal{L}_{alg}$-definable subset of $\mathbb{R}^n$ is a finite Boolean combination (i.e., finitely many intersections, unions, and complements) of sets of the form

$$\{(x_1, ..., x_n) \in \mathbb{R}^n \mid p(x_1, ..., x_n) > 0\}$$

where $p(x_1, ..., x_n)$ is a polynomial with coefficients in $\mathbb{R}$. These are called the *semialgebraic* sets. A function $f : A \subset \mathbb{R}^n \to \mathbb{R}^m$ is *semialgebraic* if its graph is a semialgebraic subset of $\mathbb{R}^n \times \mathbb{R}^m$.

As I did yesterday, I want to draw special attention to the definable sets in one variable: the $\mathcal{L}_{alg}$-definable subsets of $\mathbb{R}$. You start off with sets of the form $p(x) > 0$, where $p(x)$ is a polynomial. This is nothing other than finitely many open intervals. Consider the negation $p(x) \leq 0$; this defines finitely many closed intervals. So, the $\mathcal{L}_{alg}$-definable subsets of $\mathbb{R}$ are simply the union of finitely many intervals and points. In particular, $\mathbb{Z}$ and $\mathbb{Q}$ are not $\mathcal{L}_{alg}$-definable. This is the same as in the case of $\mathcal{L}_{lin}$, which is surprising, because the structure of $\mathcal{L}_{alg}$ seems more complex. This complexity manifests itself in higher dimensions.

As for the semilinear sets, every semialgebraic set can be written as a finite union of the intersection of finitely many sets defined by conditions of the form

$$p(x_1, ..., x_n) = 0$$
$$q(x_1, ..., x_n) > 0$$

where $p(x_1, ..., x_n)$ and $q(x_1, ..., x_n)$ are polynomials with coefficients in $\mathbb{R}$. This is just standard set theoretic manipulation of the kind I talked about yesterday.

Let me tell you now about the proof, which is the substance of what today's lecture is about. There are two steps. First we prove a geometric structure theorem that shows that any semialgebraic set can be decomposed into finitely many semialgebraic generalized cylinders and graphs. Then we deduce quantifier elimination from this. I will define what I mean by cylinders and graphs a little later on. We start with a nice result called Thom's lemma.

Lemma (Thom's Lemma [Tho65]). *Let* $p_1(X), ..., p_k(X)$ *be polynomials in the variable* X *with coefficients in* $\mathbb{R}$ *such that if* $p'_j(X) \neq 0$ *then* $p'_j(X)$ *is included among* $p_1, ..., p_k$. *Let* $S \subset \mathbb{R}$ *have the form*

$$S = \bigcap_j p_j(X) *_j 0$$

where $*_j$ is one of $<, >,$ or $=,$ then S is either empty, a point, or an open interval. Moreover, the (topological) closure of S is obtained by changing the sign conditions (changing $<$ to $\leq$ and $>$ to $\geq$).

Note: There are 3^k such possible sets, and these form a partition of $\mathbb{R}$.

Proof. The proof is by induction on k. When $k = 1$, $p_1(x)$ is a constant polynomial, so $*_j$ gives either $\mathbb{R}$ or $\emptyset$. Assume the theorem is true for $k - 1$; we will show that it must be true for k. Without loss of generality, suppose $p_k(x)$ has the largest degree of the polynomials. Then $\{p_1, ..., p_{k-1}\}$ must also satisfy the conditions of Thom's lemma. Let $*_1, ..., *_k$ be given, and form the set $S' = \bigcap_{j=1}^{k-1} p_j(X) *_j 0$. If $S' = \emptyset$ or $\{r\}$, then it clear that $S = S' \cap p_k(X) *_k 0$ has the right form. Now suppose S' is an interval I. Note that p'_k is among $p_1, ..., p_{k-1}$. On I, $p'_k(X) > 0$ or $p'_k(X) < 0$ or $p'_k(X) = 0$. So p_k is either monotone or constant on I, and so S has the right form.

We now need some "tricks" to continue with our proof. First, we identify the complex numbers $\mathbb{C}$ with $\mathbb{R}^2$ via

$$a + bi \longleftrightarrow (a, b)$$

where $a, b \in \mathbb{R}$ and $i = \sqrt{-1}$. With this identification, multiplication of complex numbers is a semialgebraic function from $\mathbb{R}^2 \times \mathbb{R}^2$ to $\mathbb{R}^2$. More generally, $\mathbb{C}^n$ is identified with $\mathbb{R}^{2n}$. Second, the collection of polynomials in the variable X with coefficients in $\mathbb{R}$ of degree not greater than n can be identified with $\mathbb{R}^{n+1}$ via

$$a_0 + a_1 X + \cdots + a_n X^n \longleftrightarrow (a_0, a_1, ..., a_n).$$

Similarly for polynomials with coefficients in $\mathbb{C}$. Addition, multiplication, differentiation of polynomials are semialgebraic functions.

Let $B_k^n(\mathbb{R})$ denote (as a subset of $\mathbb{R}^{n+1}$) the collection of polynomials in the variable X with real coefficients of degree not greater than n that have *exactly* k distinct complex roots. Let $M_k^n(\mathbb{R}) \subset B_k^n(\mathbb{R})$ be those polynomials of degree n with this property.

I'm now going to state a lemma that is vital for our proof.

Lemma ("Continuity of Roots"). *Suppose that $A \subset M_k^n(\mathbb{R})$ is connected. For each $\bar{a} \in A$ let $r_{\bar{a}}$ be the number of distinct real roots of the polynomial $p_{\bar{a}}(X)$ associated with $\bar{a}$. Then*

1. *$r_{\bar{a}} = r$ is constant on A;*
2. *There are continuous functions $f_1, ..., f_r : A \to \mathbb{R}$ such that for all $\bar{a} \in A$ we have $f_i(\bar{a}) < f_{i+1}(\bar{a})$ for $i = 1, ..., r-1$ and $p_{\bar{a}}(f_i(\bar{a})) = 0$ for $i = 1, ..., r$.*

I'm not going to prove this result, but I'll try to give you some intuition by going back to the quadratic case. Consider the polynomial $p(x) = \underline{a}x^2 + \underline{b}x + \underline{c}$. The determining factor here is the discriminant $\underline{b}^2 - 4\underline{ac}$. Consider the sets obtained by setting the discriminant equal to zero and fixing $\underline{a}$. This is a parabola in the plane $x = \underline{a}$ in $\mathbb{R}^3$ (with $\underline{b}$ and $\underline{c}$ serving as the y and z coordinates). We have zero distinct real roots below the parabola, two distinct real roots above the parabola, and one distinct real root on the parabola. So these three cases constitute connected sets.

Continuity of roots takes a little while to prove. You first prove local continuity, then use connectedness to show that the number of roots is constant across a set. This proof is involved, and I will not show it here.

Lemma. *The subsets $B^n_k(\mathbb{R})$ and $M^n_k(\mathbb{R})$ of $\mathbb{R}^{n+1}$ are semialgebraic.*

The idea here is that the polynomial $p(X)$ has a repeated root if and only if $p(X)$ and its derivative $p'(X)$ have a common factor. This can be expressed by the condition that the determinant of a matrix constructed from the coefficients of the so-called *resultant* of p and p' (also called the *discriminant* of p), has value 0. This is a semialgebraic condition on the coefficients. We can extend this idea to capture $B^n_k(\mathbb{R})$ and $M^n_k(\mathbb{R})$ semialgebraically.

I'm going to go back to the case of the quadratic equation again. Let $p(x) = ax^2 + bx + c$, so that $p'(x) = 2ax + b$. The discriminant can be written as

$$D(p, p') = \begin{vmatrix} c & b & a \\ b & 2a & 0 \\ 0 & b & 2a \end{vmatrix}$$
$$= a(4ac - b^2).$$

So we ask that $a(4ac - b^2) = 0$. This condition ensures that p has a repeated root. So we have a semialgebraic condition for p to have a repeated root.

What we've seen is that the sets $B^n_k(\mathbb{R})$ and $M^n_k(\mathbb{R})$ are semialgebraic, and as long as we stay on a connected subset of $M^n_k(\mathbb{R})$, we get the same number of real roots. And these root functions are continuous.

2.1 Graphs and Cylinders

The structure theorem shows that a semialgebraic set $S \subseteq \mathbb{R}^n$ can be partitioned into finitely many sets of two kinds, all of which are semialgebraic.

Graphs

Let $A \subset \mathbb{R}^k$ and $f : A \to \mathbb{R}$ be continuous. The *graph* of f is the subset of $\mathbb{R}^{k+1}$ given by

$$\mathrm{Graph}(f) = \{(\bar{x}, y) \in \mathbb{R}^{k+1} \mid \bar{x} \in A \text{ and } y = f(\bar{x})\}.$$

Generalized Cylinders

Let $A \subset \mathbb{R}^k$, and let $f, g : A \to \mathbb{R}$ be continuous and satisfy $f(\bar{x}) < g(\bar{x})$ for all $\bar{x} \in A$. The *cylinder* determined by f, g, and A is the subset of $\mathbb{R}^{k+1}$ given by

$$(f,g)_A = \{(\bar{x}, y) \in \mathbb{R}^{k+1} \mid \bar{x} \in A \text{ and } f(\bar{x}) < y < g(\bar{x})\}.$$

If A is connected, then graphs and cylinders based on A are connected.

Theorem (Structure Theorem). *Let S be semialgebraic. Then:*

I_n *S has finitely many connected components and each one is semialgebraic*
II_n *There is a finite partition $\mathcal{P}$ of $\mathbb{R}^{n-1}$ into connected semialgebraic sets such that for each $A \in \mathcal{P}$ there is $k_A \in \mathbb{N}$ and $f_i^A : A \to \mathbb{R} \cup \{\pm\infty\}$ for $i = 0, 1, ..., k_A + 1$ satisfying*

(a) *$f_0^A = -\infty$, $f_{k_A+1}^A = \infty$, f_i^A is continuous for $1 \leq i \leq k_A$, and $f_{i-1}^A(\bar{x}) < f_i^A(\bar{x})$ for all $1 \leq i \leq k_A + 1$ and $\bar{x} \in A$;*
(b) *all graph sets $\mathrm{Graph}(f_i^A)$ for $1 \leq i \leq k_A$ and generalized cylinders $(f_{i-1}^A, f_i^A)_A$ are semialgebraic.*
The graphs and cylinders in (b) for all $A \in \mathcal{P}$ partitions $\mathbb{R}^n$ and S.

The essence of the theorem is that S can be constructed as a finite union of semialgebraic cylinders and graphs.

Proof. The proof is by induction on n, and I shall outline the induction step. Most broadly, the argument is as follows: show $\mathrm{I}_{n-1} \Rightarrow \mathrm{II}_n$ and $\mathrm{II}_n \Rightarrow \mathrm{I}_n$. $\mathrm{II}_n \Rightarrow \mathrm{I}_n$ is evident, because the graphs are connected since the functions are semialgebraic, and the cylinders are connected since the base sets are connected. The crux is $\mathrm{I}_{n-1} \Rightarrow \mathrm{II}_n$.

Split the coordinates of $\mathbb{R}^n$ as $(x_1, ..., x_{n-1}, t)$. Using standard set theory, write S as the union of finitely many finite intersections of polynomial equalities and inequalities. Extend the finite collection of polynomials in the given definition of S by including all iterated partial derivatives with respect to t. Let this expanded list of polynomials be $q_1, ..., q_r$. The nice thing about polynomials is that we only have to do this finitely many times to obtain closure under differentiation.

For each subset $\mathcal{S} \subset \{1, ..., r\}$, form the polynomial

$$Q_{\mathcal{S}}(\bar{x}, t) = \prod_{j \in \mathcal{S}} q_j(\bar{x}, t).$$

View $\bar{x}$ as parameter variables and consider the polynomial as $Q_{\mathcal{S}, \bar{x}}(t)$, a polynomial in the variable t whose coefficients are polynomials in $\bar{x}$. For each $\ell \leq \mathrm{degree} Q_{\mathcal{S}, \bar{x}}(t)$ and $k \leq \ell$, let

$$M_{\mathcal{S}, k}^\ell = \{\bar{x} \in \mathbb{R}^{n-1} \mid \mathrm{degree} Q_{\mathcal{S}, \bar{x}}(t) = \ell \text{ and it has exactly }$$
$$k \text{ distinct real roots}\}.$$

It should come as no surprise that you can in fact show that this subspace of $\mathbb{R}^{n-1}$ is semialgebraic.

Next we partition $\mathbb{R}^{n-1}$ by taking all intersections of all $M_{S,k}^l$. This still is a finite semialgebraic partition of $\mathbb{R}^{n-1}$. Refine this partition further to obtain a partition $\mathcal{P}_0$ by taking the connected components of the sets in the partition above. By I_{n-1} this again is a finite semialgebraic partition of $\mathbb{R}^{n-1}$. For $A \in \mathcal{P}_0$, let $Q_{A,\bar{x}}(t)$ be the product of those $q_j(\bar{x},t)$ which are nonzero for (all) $\bar{x} \in A$. It can be shown that the number of roots of $Q_{A,\bar{x}}(t)$ is uniform as $\bar{x}$ ranges over A and that the 1st, 2nd,... root functions are continuous on A, as A is connected.

Form the corresponding graph and generalized cylinder sets above each set $A \in \mathcal{P}_0$. It can be shown that each such set has the form

$$\bigcap_{j=1}^{r} \{(\bar{x},t) \mid \bar{x} \in A \text{ and } q_j(\bar{x},t) *_j 0\}$$

where $*_j$ is one of $<$, $>$, or $=$. This step uses Thom's lemma.   □

That's the rough idea of the proof of the structure theorem. Now we want to use it to deduce the Tarski-Seidenberg theorem.

Theorem (Tarski–Seidenberg Theorem Redux). *Let $f : X \subset \mathbb{R}^n \to \mathbb{R}^m$ be semialgebraic. Then the image of f,*

$$f(X) = \{\bar{y} \in \mathbb{R}^m \mid \bar{y} = f(\bar{x}) \text{ for some } \bar{x} \in X\},$$

is semialgebraic.

Proof. Let $\pi : \mathbb{R}^n \times \mathbb{R}^m \to \mathbb{R}^m$ be the projection map onto the last m coordinates. Then $f(x) = \pi(\text{graph}(f))$. It thus suffices to show that the image under π of a semialgebraic set $S \subseteq \mathbb{R}^{m+n}$ is semialgebraic. We will show this by induction on n. When $n = 1$, our desired result follows directly from the structure theorem. Suppose the result is true for $n - 1$; we will show that it holds for n. Let $\pi_1 : \mathbb{R}^{n-1} \times \mathbb{R}^{1+m} \to \mathbb{R}^{1+m}$ be the projection onto the last $m + 1$ coordinates, and let $\pi_2 : \mathbb{R} \times \mathbb{R}^m \to \mathbb{R}^m$ be the projection onto the last m coordinates. Then $\pi(S) = \pi_2(\pi_1(S))$. But $\pi_1(S)$ is semialgebraic by the induction hypothesis, so $\pi(S)$ is semialgebraic by the structure theorem.   □

The Tarski–Seidenberg theorem gives quantifier elimination because if we take the formula $\exists y, \varphi(x_1, ..., x_n, y)$, where φ is a semialgebraic quantifier-free formula, then this is just the projection to $\mathbb{R}^n$ of a semialgebraic subset of $\mathbb{R}^{n+1}$, and is thus semialgebraic.

2.2 Algorithmic Cruelty

It turns out that everything I've explained can be done algorithmically. That is, we can take any definable set and algorithmically give it a semialgebraic description. The proof I've given you today relates to work done by George Collins on cylindrical algebraic decomposition. Tarski's original proof gives an algorithm for quantifier-elimination: given an $\mathcal{L}_{\text{alg}}$-formula as input, it outputs a quantifier-free formula that defines the same set as the input formula. Computational efficiency of a quantifier elimination algorithm thus becomes important for applications (e.g., robot motion planning). Cylindrical algebraic decomposition-based quantifier elimination, such as described above and developed in 1975 by Collins [Col75], has played an important role.

Quantifier elimination for $\mathbb{R}_{\text{alg}}$ is, unfortunately, an inherently computationally intensive problem. It is known that there is a doubly exponential lower bound in the number of quantifiers for worst-case time complexity. So, quantifier elimination is something that is do-able in principle, but not by any computer that you and I are ever likely to see. Well, I'll retract that last statement because it's probably false.

3 *O*-minimal Structures

An $\mathcal{L}$-structure $\mathfrak{R}$ is *o-minimal* if every definable subset of $\mathbb{R}$ is the union of finitely many points and open intervals (a, b), where $a < b$ and $a, b \in \mathbb{R} \cup \{\pm\infty\}$. Thus far we have seen two examples of *o*-minimal structures: $\mathbb{R}_{\text{lin}}$, the semilinear context, and $\mathbb{R}_{\text{alg}}$, the semialgebraic context. We showed by quantifier elimination that these structures are *o*-minimal. *O*-minimal is short for ordered minimal. We use this name because, for an *o*-minimal structure, the definable subsets of $\mathbb{R}$ are exactly those that must be there because of the presence of $<$. The hypothesis of *o*-minimality combined with the power of definability have remarkable consequences.

3.1 Minimal Structures

Consider the field of complex numbers $(\mathbb{C}, \cdot, +)$. There is a theorem due to Chevalley which says that $(\mathbb{C}, \cdot, +)$ has the quantifier elimination property. All definable subsets of $\mathbb{C}$ can be defined by finite Boolean combinations of polynomials equalities $p(x) = 0$ or inequalities $p(x) \neq 0$, where p has coefficients in $\mathbb{C}$. As long as p is not the zero polynomial, $p(x)$ will consist of finitely many points, and $p(x) \neq 0$ will consist of cofinitely many points. So all definable subsets of $\mathbb{C}$ are either finite or cofinite. Notice that these are the sets you *have* to be able to define using equality. So $(\mathbb{C}, \cdot, +)$ is minimal.

Minimal structures were first know by model theorists. One of my contributions was to look at the properties of minimal structures when we have an ordering. The theme of what I'm going to be talking about today and in the remaining lectures is *o*-minimal structures.

3.2 O-minimal Structures

The first result, which I'm going to spend quite a bit of time on, is one of the most important results on o-minimality.

Theorem (Monotonicity Theorem). *Let $\mathfrak{R}$ be an $\mathcal{L}$-structure that is o-minimal. Suppose that $f : \mathbb{R} \to \mathbb{R}$ is $\mathfrak{R}$-definable. Then there are $-\infty = a_0 < a_1 < \cdots < a_{k-1} < a_k = \infty$ in $\mathbb{R} \cup \{\pm\infty\}$ such that for each $j < k$ either $f \upharpoonright (a_j, a_{j+1})$ is constant or is a strictly monotone bijection of (possibly unbounded) open intervals in $\mathbb{R}$.*

In particular, all definable $f : \mathbb{R} \to \mathbb{R}$ are piecewise continuous. Now we can see the power of o-minimality. We are not confined to looking at semialgebraic sets.

I'm going to give you some idea of what's in the proof of the monotonicity theorem. We need to show that for every open interval $I \subseteq \mathbb{R}$, there is an open interval $I^* \subseteq I$ on which f is constant or strictly montone. This will give us a local version of the theorem, which we need to globalize. For now, assume the local version and consider the task of globalization.

Let $\theta(x)$ say "x is the left endpoint of an interval on which f is constant or strictly monotone and this interval cannot be extended properly on the left while preserving this property." You should check that this is a meaningful expression in the context of our structure. Clearly $\theta(x)$ defines a finite subset of $\mathbb{R}$, say $b_1, ..., b_p$. We know that $f \upharpoonright (b_i, b_{i+1})$ is strictly monotone or constant, because otherwise the local version of the theorem implies that $(b_i, b_{i+1}) \cap \theta(x) \neq \emptyset$. To finish working from local to global, we note that if $f \upharpoonright (b_i, b_{i+1})$ is strictly monotone or constant then $f((b_i, b_{i+1}))$ consists of finitely many points and intervals, and so we can partition each (b_i, b_{i+1}) into a finite collection of points and intervals in such a way that the image under f of each interval in the partition is itself an open interval.

Now I'll try to give you some idea of what's in the proof of the local version of the theorem. Define the following formulas:

$\varphi_0(x)$ says "$\exists$ open interval J with x as an endpoint on which f is constantly equal to $f(x)$."

$\varphi_1(x)$ says "$\exists$ open interval J containing x such that $f(y) < f(x)$ for $y \in J$, $y < x$ and $f(y) > f(x)$ for $y \in J$, $y > x$."

$\varphi_2(x)$ says "$\exists$ open interval J containing x such that $f(y) < f(x)$ for $y \in J$, $y > x$ and $f(y) > f(x)$ for $y \in J$, $y < x$."

$\varphi_3(x)$ says "$\exists$ open interval J containing x such that $f(y) < f(x)$ for $y \in J$, $y \neq x$."

$\varphi_4(x)$ says "$\exists$ open interval J containing x such that $f(y) > f(x)$ for $y \in J$, $y \neq x$."

These five formulas exhaust all possibilities, by o-minimality. Also by o-minimality, possibilities 3 and 4 cannot hold, so $\exists I_1 \subseteq I$ contained in the set defined by one of $\varphi_0, ..., \varphi_2$. If I_1 is contained in the set defined by φ_0, then f is constant on I_1. If I_1 is contained in the set defined by φ_1, then f is strictly monotone increasing on I_1. And if I_1 is contained in the set defined by φ_2, then f is strictly monotone decreasing on I_1.

3.3 Cells

What I want to do next is talk about a particular kind of definable set: cells. Cells in $\mathbb{R}$ are either points (0-cells) or intervals (1-cells). Cells in $\mathbb{R}^2$ can be constructed as follows: begin with cells in $\mathbb{R}$, and take the graph of a continuous function defined on those cells, as well as the cylinders defined by pairs of continuous functions.

More formally, let $\mathfrak{R}$ be an $\mathcal{L}$-structure. The collection of $\mathfrak{R}$-cells is a subcollection $\mathcal{C} = \bigcup_{n=1}^{\infty} \mathcal{C}_n$ of the $\mathfrak{R}$-definable subsets of $\mathbb{R}^n$ for $n = 1, 2, 3, ...$ defined recursively as follows.

Cells in $\mathbb{R}$

The collection of cells $\mathcal{C}_1$ in $\mathbb{R}$ consists of all single point sets $\{r\} \subset \mathbb{R}$ and all open intervals $(a, b) \subseteq \mathbb{R}$, where $a < b$ and $a, b \in \mathbb{R} \cup \{\pm\infty\}$.

Cells in $\mathbb{R}^{n+1}$

Assume the collection of cells $\mathcal{C}_n$ in $\mathbb{R}_n$ have been defined. The collection $\mathcal{C}_{n+1}$ of cells in $\mathbb{R}_{n+1}$ consist of two different kinds: graphs and cylinders.

Graphs

Let $C \in \mathcal{C}_n$ and let $f : C \subseteq \mathbb{R}^n \to \mathbb{R}$ be $\mathfrak{R}$-definable and continuous. Then $\mathrm{Graph}(f) \subseteq \mathbb{R}^{n+1}$ is a cell.

Generalized Cylinders

Let $C \in \mathcal{C}_n$. Let $f, g : C \subseteq \mathbb{R}^n \to \mathbb{R}$ be $\mathfrak{R}$-definable and continuous such that $f(\bar{x}) < g(\bar{x})$ for all $\bar{x} \in C$. Then the cyclinder set $(f, g)_C \subset \mathbb{R}^{n+1}$ is a cell.

Cells are $\mathfrak{R}$-definable and connected. There is a concept of dimension for cells: for each cell $C \subseteq \mathbb{R}^n$ there is a largest $k \leq n$ and $i_1, ..., i_k \in \{1, 2, ..., n\}$ such that if $\pi : \mathbb{R}^n \to \mathbb{R}^k$ is the projection mapping given by $\pi(x_1, ..., x_n) = (x_{i_1}, ..., x_{i_k})$, then $\pi(C) \subseteq \mathbb{R}^k$ is an open cell in $\mathbb{R}^k$. This value of k we call the *dimension* of C. Basically speaking, the dimension of a cell is the number of times we construct cylinders in the "bottom-up" construction of a cell.

3.4 Cell Decomposition

An $\mathfrak{R}$-*decomposition* $\mathcal{D}$ of $\mathbb{R}^n$ is a partition of $\mathbb{R}^n$ into finitely many $\mathfrak{R}$-cells satisfying:

1. If $n = 1$, then $\mathcal{D}$ consists of finitely many open intervals and points.
2. If $n > 1$ and $\pi_n : \mathbb{R}^n \to \mathbb{R}^{n-1}$ denotes projection onto the first $n-1$ coordinates, then $\{\pi_n(C) : C \in \mathcal{D}\}$ is a decomposition of $\mathbb{R}^{n-1}$.

This is just a generalized version of the cylindrical algebraic decomposition.

Theorem (Cell Decomposition Theorem). *Let $\mathfrak{R}$ be o-minimal and let $S \subset \mathbb{R}^n$ be definable. Then there is a decomposition $\mathcal{D}$ of $\mathbb{R}^n$ that partitions S into finitely many cells. In particular, if $f : A \subset \mathbb{R}^n \to \mathbb{R}$ is definable, then there is a partition of A into cells such that the restriction of f to each cell is continuous.*

Some Obvious Consequences

- Using the definition of a cell as defined above, we obtain a good geometric definition of the dimension of a definable set. Namely, the dimension of a set is the maximimum dimension of the cells in a decomposition that partitions it.
- Since cells are connected, it follows that every definable set has finitely many connected components. This is one aspect of the "tameness" alluded to in the name "tame topology."
- The topological closure of a definable set consists of finitely many connected components; same for the the interior and the frontier (or boundary). Even if you start with a connected set, this won't necessarily be true unless the set is definable. Consider a comb with infinitely many teeth, where the points of those teeth are not in the set; the frontier of this connected set consists of infinitely many connected components.

3.5 Definable Families

Let $S \subset \mathbb{R}^{n+p}$ be a definable set in the o-minimal structure $\mathfrak{R}$. For each $\bar{b} \in \mathbb{R}^n$ define $S_{\bar{b}} := \{\bar{y} \in \mathbb{R}^p \mid (\bar{b}, \bar{y}) \in S\}$. (Note that some $S_{\bar{b}}$ may be empty.) The family $\{S_{\bar{b}} \mid \bar{b} \in \mathbb{R}^n\}$ of subsets of $\mathbb{R}^p$ is called a *definable family*.

The next result is quite surprising; we did not expect to find it.

Theorem (Uniform Bounds Theorem). *Let $\mathfrak{R}$ be o-minimal and let $S \subset \mathbb{R}^{n+p}$ be a definable set such that $S_{\bar{b}}$ is finite for all $\bar{b} \in \mathbb{R}^n$. Then there is a fixed $K \in \mathbb{N}$ satisfying $|S_{\bar{b}}| \leq K$ for all $\bar{b} \in M^n$.*

Note that a definable subset of $\mathbb{R}$ in an *o*-minimal structure $\mathfrak{R}$ is infinite if and only if it contains an interval. So the theorem actually is stronger. It can be thought of as a generalization of our earlier result on the roots of polynomial equations.

Recall our discussion of minimality and the field of complex numbers. We can also obtain a version of the uniform bounds theorem for subsets of $\mathbb{C}^{n+1}$.

3.6 Sketch Proof of the Cell Decomposition Theorem

Next I want to give you a quick sketch of the proof of the cell decomposition theorem. The proof is by induction on n. Define the following induction hypotheses:

$\mathbf{I}_n$ There exists a cell decomposition that respects finitely many definable subsets of $\mathbb{R}^n$.

$\mathbf{II}_n$ Let $f : A \subseteq \mathbb{R}^n \to \mathbb{R}$ be definable. Then there is a decomposition $\mathcal{D}$ of $\mathbb{R}^n$ that partitions A such that $f \upharpoonright C$ is continuous for all $C \in \mathcal{D}, C \subseteq A$.

$\mathbf{III}_n$ The uniform finiteness property holds for all definable families $\{S_{\bar{b}} \mid \bar{b} \in \mathbb{R}^n\}$, where $S \subseteq \mathbb{R}^{n+p}$ is definable.

$\mathbf{I}_1$ is true by the definition of *o*-minimality, $\mathbf{II}_1$ is just the monotonicity theorem, and $\mathbf{III}_1$ requires an intricate direct argument that we will not go into. By assuming $\mathbf{I}_m, \mathbf{II}_m, \mathbf{III}_m$ for all $m < n$, we can use the monotonicity theorem to show that $\mathbf{I}_n$ holds, and then that $\mathbf{II}_n$ holds, and finally that $\mathbf{III}_n$ holds.

3.7 Quantifier Elimination

Theorem (van den Dries [Van98]). *Let I be an index set and for each $i \in I$ let $f_i : \mathbb{R}^{n_i} \to \mathbb{R}$ be (total) analytic functions. Then the structure $(\mathbb{R}_{\mathrm{alg}}, \{f_i : i \in I\})$ admits quantifier elimination if and only if each f_i is semialgebraic.*

So, e.g., $\mathbb{R}_{\exp}$, the real exponential field does not have quantifier elimination. Quantifier elimination is not easy to come by.

3.8 Partial Elimination

Suppose that a structure $\mathfrak{R}$ has the property that every definable set is definable by an *existential formula*, that is, a formula having the form

$$\exists x_1 \exists x_2 \cdots \exists x_k \, \varphi$$

where φ is a quantifier-free $\mathcal{L}$-formula. How can this help? Suppose that the $\mathfrak{R}$-definable sets that are definable using quantifier-free formulas can be analyzed, and that all such have finitely many connected components. The continuous image of a connected set is connected (elementary topology). Existential quantification corresponds to projection, and projection is a continuous map. Thus all $\mathfrak{R}$-definable subsets of $\mathbb{R}$ have finitely many connected components, that is, all such are the union of finitely many points and open intervals. We conclude that $\mathfrak{R}$ is o-minimal, and so all the geometric and topological properties available as consequences of o-minimality apply. We will talk more about this tomorrow. Tomorrow I want to give you a deeper understanding of what tame topology means, and take you on a tour of structures that we now know to be o-minimal.

4 Examples and Some Further Properties

I'm going to start off by giving you a tour of some o-minimal structures, and I'll finish today by giving you some of the finer tame topological results.

4.1 Examples of O-minimal Structures

So far we have two examples of o-minimal structures: $\mathbb{R}_{\text{lin}}$, the semilinear context, and $\mathbb{R}_{\text{alg}}$, the semialgebraic context. As I shall discuss later, o-minimality implies a wealth of good analytic and topological properties. This provided ample motivation to seek out o-minimal structures that expand $\mathbb{R}_{\text{alg}}$ to include transcendental data.

We now survey some of the remarkable results that have been obtained beginning in the mid-1980s.

4.2 Expansions of $\mathbb{R}_{\text{alg}}$

Consider the class of *restricted analytic functions*, $\vec{an}$, where $g : \mathbb{R}^n \to \mathbb{R} \in \vec{an}$ if there is some analytic $f : U^{\text{open}} \subset \mathbb{R}^n \to \mathbb{R}$ such that $[0,1]^n \subset U$, $g \restriction [0,1]^n = f \restriction [0,1]^n$, and $g(\bar{x}) = 0$ otherwise. (The point of doing something like this is that you can avoid considering behaviour at $\pm\infty$ by appropriate restriction of the functions' domain.) Let $\mathbb{R}_{\text{an}}$ be the expansion of $\mathbb{R}_{\text{alg}}$ by adding as basic functions all $g \in$ an. Then the structure $\mathbb{R}_{\text{an}}$ admits elimination down to existential formulas and is o-minimal (van den Dries [Van86]).

Let me give you a sense of what kind of functions you can actually get here. You can not only get bounded functions, but also functions that live on all of $\mathbb{R}$. For instance, we can obtain $\arctan x$ from the restriction of $\sin x$ and $\cos x$ to $(-\pi, \pi)$. Everything I said about definable functions—monotonicity and the cell decomposition—works for these functions. This result depends on the work of Lojasiewicz and Gabrielov (in the 1960s). Another useful fact about $\mathbb{R}_{\text{an}}$ is that it is polynomially bounded.

Polynomial growth: Let $f : (a, \infty) \to \mathbb{R}$ be definable in $\mathcal{R}_{\text{an}}$. Then there is some $N \in \mathbb{N}$ such that $|f(x)| < x^N$ for sufficiently large x.

Adjoin to $\mathbb{R}_{\text{an}}$ the function $^{-1}$ given by $x \longmapsto 1/x$ for $x \neq 0$ and $0^{-1} = 0$.

Theorem (Denef–van den Dries [DV88]). ($\mathbb{R}_{\text{an}}, ^{-1}$) *admits elimination of quantifiers.*

Whereas in the semialgebraic context we know all the basic functions (they are just polynomials), in this case we have a much larger collection of basic functions, so that our descriptive language is much richer. The languages in (1) and (2) are large, but nonetheless natural. Quantifier elimination always can be achieved by enlarging the language, but no advantage is gained: in general, the quantifier-free sets thus obtained can be horribly badly behaved.

The next theorem is really quite spectacular, and was a breakthrough for the subject.

Theorem (Wilkie [Wil96]). $\mathbb{R}_{\text{exp}}$ *admits elimination down to existential formulas.*

O-minimality then follows by a result of Khovanskii [Kho80] (which Wilkie also uses in his proof). Recall that yesterday I showed that quantifier elimination is not possible in $\mathbb{R}_{\text{exp}}$. This result tells us "the best that we can do." This was the first *o*-minimal structure with functions growing faster than polynomials.

This theorem addresses a question posed originally by Tarski. He asked if his results on $\mathbb{R}_{\text{alg}}$ could be extended to $\mathbb{R}_{\text{exp}}$. Wilkie's result from the syntactic and topological points of view is the best possible.

Macintyre and Wilkie [MW96] link decidability of the theory of the real exponential field to the following conjecture.

Conjecture (Schanuel's Conjecture [Sch91]). *Let* $r_1, ..., r_n \in \mathbb{R}$ *be linearly independent over* $\mathbb{Q}$. *Then the transcendence degree over* $\mathbb{Q}$ *of* $\mathbb{Q}(r_1, ..., r_n, e^{r_1}, ..., e^{r_n})$ *is at least* n.

This has implications about the transcendence of various things, which I might talk more about tomorrow. Schanuel's conjecture is now regarded by mathematicians as being intractable, so I don't know if we will ever see it verified. But most mathematicians seem to believe that it is true.

Theorem (Macintyre–Wilkie [MW96]). *Schanuel's conjecture implies that the theory of the real exponential field is decidable.*

A natural question is to ask what happens if we combine the restricted analytic and the exponential functions in our basic functions. Van den Dries and Miller adapt Wilkie's techniques to prove the following.

Theorem (van den Dries–Miller [VM94]). $\mathbb{R}_{\text{an,exp}}$ *admits elimination down to existential formulas and (by Khovanskii) is o-minimal.*

Inspired by the work of Ressayre [Res93], these authors analyze $\mathbb{R}_{\text{an,exp}}$ further.

Theorem (van den Dries–Macintyre–Marker [VMM94]). $\mathbb{R}_{\mathrm{an,exp,log}}$ *admits elimination of quantifiers.*

Their analysis further shows that every definable function in one variable is bounded by an iterated exponential. Macintyre–Marker [MM97] show that the logarithm is necessary for the quantifier elimination. In a second paper van den Dries–Macintyre–Marker [VMM97] develop tools that enable them to obtain several further results.

What I want to do now is talk about results of definability and undefinability. Let $f(x) = (\log x)(\log\log x)$ and let $g(x)$ be a compositional inverse to f defined on some interval (a, ∞). Hardy [Har12] conjectured in 1912 that g is not asymptotic to a composition of exp, log, and semialgebraic functions.

Theorem (van den Dries–Macintyre–Marker [VMM97]). *Hardy's conjecture is true.*

Let me talk about some other undefinability results which people will perhaps find more down to earth. Building on some remarkable ideas and results of Mourges–Ressayre [MR93], van den Dries–Macintyre–Marker derive some "undefinability" results also.

Theorem (van den Dries–Macintyre–Marker [VMM97]). *None of the following functions is definable in* $\mathbb{R}_{\mathrm{an,exp}}$.

 i. *the restriction of the gamma function* $\Gamma(x) = \int_0^\infty e^{-t}t^{x-1}\,dt$ *to* $(0, \infty)$,
 ii. *the error function* $\int_0^x e^{-t^2}\,dt$,
 iii. *the logarithmic integral* $\int_x^\infty t^{-1}e^{-t}\,dt$,
 iv. *the restriction of the Riemann zeta function* $\zeta(s) = \sum_{n=1}^\infty n^{-s}$ *to* $(1, \infty)$.

These are functions that we deal with all the time, but are not definable in the context of $\mathbb{R}_{\mathrm{an,exp}}$. So our work is not yet complete.

For $r \in \mathbb{R}$ let x^r denote the real power function

$$x^r = \begin{cases} x^r & \text{if } x > 0 \\ 0 & \text{if } x \le 0. \end{cases}$$

Let $\mathbb{R}_{\mathrm{an}}^{\mathbb{R}}$ denote the expansion of $\mathbb{R}_{\mathrm{an}}$ by all power functions x^r for $r \in \mathbb{R}$.

Theorem (Miller [Mil94]). $\mathbb{R}_{\mathrm{an}}^{\mathbb{R}}$ *has elimination of quantifiers.*

Again, this is a structure with functions whose growth at infinity is bounded by polynomials. An expansion $\mathfrak{R}$ of $\mathbb{R}_{\mathrm{alg}}$ is *polynomially bounded* if for every definable $f : \mathbb{R} \to \mathbb{R}$ there is some $N \in \mathbb{N}$ so that $|f(x)| < x^N$ for sufficiently large x. I've given you examples of functions that are polynomially bounded (an) and those that are not (exp).

Theorem (Growth Dichotomy Theorem). *Let $\mathfrak{R}$ be an o-minimal expansion of $\mathbb{R}_{\mathrm{alg}}$. Then either the exponential function e^x is definable in $\mathfrak{R}$ or $\mathfrak{R}$ is polynomially bounded. In the second case, for every definable $f : \mathbb{R} \to \mathbb{R}$ in $\mathfrak{R}$ not ultimately identically zero, there are $c \in \mathbb{R}\backslash\{0\}$ and $r \in \mathbb{R}$ such that $f(x) = cx^r + o(x^r)$ as $x \to \infty$.*

This amazing theorem shows that there is no "middle ground": if functions are not bounded by polynomial growth, then exponential growth must be possible.

Let $\partial\Phi$ be a collection of restricted analytic functions that is closed under differentiation. Since derivatives are definable in $\mathbb{R}_{\mathrm{an}}$ (definability of derivatives was mentioned in Section 1), all of the functions in $\partial\Phi$ are definable in this structure.

Theorem. *The structure $(\mathbb{R}_{\mathrm{alg}}, f)_{f\in\partial\Phi}$ has elimination down to existential formulas.*

This result does not give us new functions, but tells us how we can think of what we have in a nicer way. The next result gives us new functions.

Using (delicate) generalized power series methods new expansions of $\mathbb{R}_{\mathrm{alg}}$ are constructed in van den Dries–Speissegger [VS98]. There are two polynomially bounded versions that have elimination down to existential formulas: generalized convergent power series (using real, rather than integer, powers) and multisummable series. Moreover, the exponential function can be added while preserving *o*-minimality. If, in addition the logarithmic function is adjoined as a basic function, these expansions admit quantifier elimination. In one of these expansions, the gamma function on $(0, \infty)$ is definable, and in the second, the Riemann zeta function on $(1, \infty)$ is definable.

Now we come to some really beautiful results of Wilkie [Wil96]. I have to introduce another class of functions, which again is quite natural. A function $f : \mathbb{R}^n \to \mathbb{R}$ is said to be *Pfaffian* if there are functions $f_1, ..., f_k : \mathbb{R}^n \to \mathbb{R}$ and polynomials $p_{ij} : \mathbb{R}^{n+i} \to \mathbb{R}$ such that

$$\frac{\partial f_i}{\partial x_j}(\bar{x}) = p_{ij}(\bar{x}, f_1(\bar{x}), ..., f_i(\bar{x}))$$

for all $i = 1, ..., k$, $j = 1, ..., n$, and $\bar{x} \in \mathbb{R}^n$. So this is a chaining procedure, going one step at a time to generate more complicated functions. Wilkie proved (by quite different methods than used previously) that the expansion of $\mathbb{R}_{\mathrm{alg}}$ by all Pfaffian functions is *o*-minimal. The introduction of Pfaffian functions allows us to integrate and retain *o*-minimality. Our class of functions is now even richer.

Speissegger [Sep99] extends Wilkie's methods to obtain the "Pfaffian closure" of an *o*-minimal expansion of $\mathbb{R}_{\mathrm{alg}}$. In particular, such a structure is closed under integration (antidifferentiation) of functions in one variable.

I have one more comment before I move on to the second part of this talk. What about quantifier elimination in these expansions? Unfortunately, this is completely unknown. We have seen that many geometric results are obtainable for these o-minimal strcutures, but results on quantifier elimination are not available.

4.3 Finer Analytic and Topological Consequences of O-minimality

For this section, assume throughout that we work in some o-minimal expansion $\mathfrak{R}$ of $\mathbb{R}_{\mathrm{alg}}$. We showed earlier in our discussion of the cell decomposition that our functions are continuous on each cell. A natural question to ask is: can we do better than this? The next theorem shows that the answer is yes.

Theorem ($\mathcal{C}^k$ Cell Decomposition Theorem). *For each definable set $X \subset \mathbb{R}^m$ and $k = 1, 2, ...$, there is a decomposition of $\mathbb{R}^m$ that respects X and for which the data in the decomposition are $\mathcal{C}^k$.*

In the next theorem, "definably homeomorphic" means that the homeomorphism between structures is itself a definable function.

Theorem (Triangulation Theorem). *Every definable set $X \subset \mathbb{R}^m$ is definably homeomorphic to a semilinear set. More precisely, X is definably homeomorphic to a union of simplices of a finite simplicial complex in $\mathbb{R}^m$.*

Note that these results are "nice" topological results, in the spirit of the term "tame topology" coined by Grothendieck. The next theorem is another nice result. It says that if we look at the fibers of some set, then the fibers corresponding to a given connected component of that set are homeomorphic.

Theorem (Number of Homeomorphism Types). *Let $S \subset \mathbb{R}^{m+n}$ be definable, so that $\{S_{\bar{a}} \mid \bar{a} \in \mathbb{R}^m\}$ is a definable family of subsets of $\mathbb{R}^n$. Then there is a definable partition $\{B_1, ..., B_p\}$ of $\mathbb{R}^m$ such that for all $\bar{a}_1, \bar{a}_2 \in \mathbb{R}^m$, the sets $S_{\bar{a}_1}$ and $S_{\bar{a}_2}$ are homeomorphic if and only if there is some $j = 1, ..., p$ such that $\bar{a}_1, \bar{a}_2 \in B_j$.*

Uniform finiteness combined with Wilkie's theorem yields Khovanskii's theorem. You all know the theorem that says that a polynomial of degree k has no more than k distinct real roots. One of the implications of Khovanskii's theorem is that there is a uniform bound on the number of distinct real roots of $p_{l,k}(x) = ax^l + bx^k$ as l, k vary.

Theorem (Khovanskii [Kho91]). *There exists a bound in terms of m and n for the number of connected components of a system of n polynomial inequalities with no more than m monomials.*

There is a trick to proving this theorem. Replace x^m by $e^{m \log x}$, and let m vary over $\mathbb{R}$. The set of (a, b, m, n, x) such that $ae^{m \log x} + be^{n \log x} = 0$ is in $\mathbb{R}_{\exp}$. Now fix x and let the other parameters vary. Uniform finiteness gives us a bound on the size of the fibers.

The next theorem gives an o-minimal improvement of the previous result.

Theorem. *There is a bound in terms of m and n for the number of homeomorphism types of the zero sets in $\mathbb{R}^n$ of polynomials $p(x_1, ..., x_n)$ over $\mathbb{R}$ with no more than m monomials.*

Theorem (Marker–Steinhorn [MS94]). *Let $\mathfrak{R}$ be an o-minimal expansion of $\mathbb{R}_{\mathrm{alg}}$, and let $g_{\bar{a}} : B \subseteq \mathbb{R}^m \to \mathbb{R}$ for $\bar{a} \in A \subset \mathbb{R}^m$ be an $\mathfrak{R}$-definable family $\mathcal{G}$ of functions. Then every $f : B \subseteq \mathbb{R}^m \to \mathbb{R}$ which is in the closure of $\mathcal{G}$ is definable in $\mathfrak{R}$.*

Here, closure refers to closure in the product topology $\mathbb{R}^B$. This result is very surprising. Consider it in the case of semialgebraic functions: the pointwise limit of semialgebraic functions is semialgebraic.

4.4 The Euler Characteristic

Now what I want to talk about is how we can begin to do algebraic topology from the viewpoint of o-minimality. We would like to consider the Euler characteristic in an o-minimal context.

Let $S \subset \mathbb{R}^n$ be definable and $\mathcal{P}$ be a partition of S into cells. Let $n(\mathcal{P}, k)$ be the number of cells of dimension k in $\mathcal{P}$, and define $E_{\mathcal{P}}(S) = \sum (-1)^k n(\mathcal{P}, k)$.

Proposition. *If $\mathcal{P}$ and $\mathcal{P}'$ are partitions of S into cells, then $E_{\mathcal{P}}(S) = E_{\mathcal{P}'}(S)$.*

So we define $E(S) = E_{\mathcal{P}}(S)$ for any partition $\mathcal{P}$. The Euler characteristic has played a very interesting role in o-minimal theory generally. Its properties include the following:

1. Let A and B be disjoint definable subsets of $\mathbb{R}^n$. Then $E(A \cup B) = E(A) + E(B)$.
2. Let $A \subset \mathbb{R}^m$ and $B \subset \mathbb{R}^n$ be definable. Then $E(A \times B) = E(A)E(B)$.
3. Let $f : A \subset \mathbb{R}^m \to \mathbb{R}^n$ be definable and injective. Then $E(A) = E(f(A))$.

5 VC Dimension and Applications

Before I begin, I want to thank the organizer, Don. I know it was talked about on Wednesday night—the wonderful attention you pay to graduate students—and I can see that this week. And for me too, not a graduate student, it has been a wonderful week. So I just want to thank Don for that.

5.1 Vapnik-Červonenkis Dimension

A collection $\mathcal{C}$ of subsets of a set X *shatters* a finite subset F if $\{F \cap C \mid C \in \mathcal{C}\} = \mathcal{P}(F)$, where $\mathcal{P}(F)$ is the set of all subsets of F. The collection $\mathcal{C}$ is a

VC-class if there is some $n \in \mathbb{N}$ such that no set F containing n elements is shattered by $\mathcal{C}$, and the least such n is the VC-dimension, $\mathcal{V}(\mathcal{C})$, of $\mathcal{C}$.

Let $\mathcal{C} \cap F := \{\mathcal{C} \cap F \,|\, C \in \mathcal{C}\}$ and for $n = 1, 2, ...$, let

$$f_{\mathcal{C}}(n) := \max\{|\mathcal{C} \cap F| \,|\, F \subset X \text{ and } |F| = n\}.$$

Also, let $p_d(n) = \sum_{i<d} \binom{n}{i}$. The next theorem gives us a polynomial bound on the growth of $f_{\mathcal{C}}$ for VC-classes.

Theorem (Sauer [Sau72]). *Suppose that $f_{\mathcal{C}}(d) < 2^d$ for some d. Then $f_{\mathcal{C}}(n) \leq p_d(n)$ for all n.*

To prove Sauer's theorem we need the following proposition.

Proposition. *Let $|F| = n$, and let $\mathcal{D}$ be a collection of subsets of F such that $|\mathcal{D}| > p_d(n)$, where $d \leq n$. Then there is $E \subseteq F$, $|E| = d$, so that $\mathcal{D}$ shatters E.*

Proof. The proof is by induction on n. The result is clear when $d = 0$ or $d = n$, so assume $0 < d < n$. Fix $x \in F$ and let $F' = F\backslash\{x\}$, $\mathcal{D}' = \{D\backslash\{x\} \,|\, D \in \mathcal{D}\}$. Consider the map $\pi(D) = D\backslash\{x\}$. Note that $\pi^{-1}(D')$ has either one or two elements (i.e., D', $D' \cup \{x\}$), depending on whether or not $x \in D$. Write $\mathcal{D}' = \mathcal{D}'_1 \cup \mathcal{D}'_2$, where $\mathcal{D}'_1$ is the class of all sets with one preimage under π, and $\mathcal{D}'_2$ is the class of all sets with two preimages under π. If $|\mathcal{D}'| > p_d(n-1)$, then by the induction hypothesis we have $E' \subseteq F'$, $|E'| = d$, so that $\mathcal{D}'$ shatters E'. It follows that $\mathcal{D}$ shatters E'. If $|\mathcal{D}'| \leq p_d(n-1)$, then we have $|\mathcal{D}| = |\mathcal{D}'_1| + 2|\mathcal{D}'_2| = |\mathcal{D}'| + |\mathcal{D}'_2|$. But $|\mathcal{D}| > p_d(n) = p_d(n-1) + p_{d-1}(n-1)$, and so $|\mathcal{D}'_2| > p_{d-1}(n-1)$. Thus by the induction hypothesis we have $E' \subseteq F'$, $|E'| = d - 1$ so that $\mathcal{D}'_2$ shatters E'. It follows that $\mathcal{D}$ shatters $E' \cup \{x\}$. $\square$

Now we can use our proposition to prove Sauer's theorem.

Proof. If $d > n$ then $p_d(n) = 2^n$, and the inequality holds trivially. So let $d \leq n$. Consider an arbitrary set $F \subset X$ with $|F| = n$. If $|\mathcal{C} \cap F| > p_d(n)$, then by our proposition there exists $E \subseteq F$, $|E| = d$ such that $\mathcal{C}$ shatters E. But this contradicts $f_{\mathcal{C}}(d) < 2^d$. Thus for all F we must have $|\mathcal{C} \cap F| \leq p_d(n)$, implying that $f_{\mathcal{C}}(d) \leq p_d(n)$. $\square$

Now I'm going to try to connect back to the logical issues we have been discussing through the week. An $\mathcal{L}$-formula $\varphi(x_1, ..., x_k; y_1, ..., y_m)$ has the *independence property* with respect to the $\mathcal{L}$-structure $\mathfrak{R}$ if for every $n = 1, 2, ...$, there are $\bar{b}_1, ..., \bar{b}_n \in \mathbb{R}^m$ such that for every $X \subseteq \{1, ..., n\}$, there is some $\bar{a}_X \in \mathbb{R}^k$ satisfying

$$\varphi(\bar{a}_X; \bar{b}_i) \text{ is true in } \mathfrak{R} \Leftrightarrow i \in X.$$

If φ does not have the independence property with respect to $\mathfrak{R}$, we let $\mathcal{I}(\varphi)$ be the least n for which the property above fails.

For an $\mathcal{L}$-formula $\varphi(\bar{x}; \bar{y})$ and a structure $\mathfrak{R}$, let $S \subseteq \mathbb{R}^{k+m}$ be the set defined by φ. We let $C_\varphi := \{S_{\bar{b}} \mid \bar{b} \in \mathbb{R}^m\}$ denote the family of subsets of $\mathbb{R}^k$ determined by S.

Now what I want to do is come to the connection between the concept I have just defined, and VC dimension.

Theorem (Laskowski [Las92a]). *The definable family C_φ is a VC-class if and only if φ does not have the independence property. Moreover, if $\mathcal{V}(C_\varphi) = d$ and $\mathcal{I}(\varphi) = n$, then $n \leq 2^d$ and $d \leq 2^n$ (and these bounds are sharp).*

Let $\psi(\bar{y}; \bar{x}) := \varphi(\bar{x}; \bar{y})$ be the *dual formula* of φ. That is, ψ and φ are the same formula (and so define the same set) with the roles of $\bar{x}$ and $\bar{y}$ reversed. The theorem follows from the next two lemmas.

Lemma 1. *With the notation as above, $\mathcal{V}(C_\varphi) = d$ if and only if $\mathcal{I}(\psi) \geq d$.*

Proof. By definition, $\mathcal{V}(C_\varphi) \geq d$ if and only if there exist $\bar{a}_1, \bar{a}_2, ..., \bar{a}_d \in \mathbb{R}^k$ such that for every $X \subseteq \{1, ..., d\}$ there is $b_X \in \mathbb{R}^m$ for which $\varphi(\bar{a}_j, \bar{b}_X)$ true $\Leftrightarrow j \in X$. This exactly says: $\mathcal{I}(\psi) \geq d$. $\qquad\square$

Understanding Lemma 1 is just a matter of understanding the definitions of VC dimension and the independence number.

Lemma 2. *Let the notation be as above. Then $\mathcal{I}(\varphi) = n$ implies $\mathcal{I}(\psi) \leq 2^n$.*

Proof. Suppose $\mathcal{I}(\psi) > 2^n$. By definition, there are $\bar{b}_s \in \mathbb{R}^k$ for each $s \subseteq \{1, ..., n\}$ so that for every $X \subseteq \{s \mid s \subseteq \{1, ..., n\}\}$ there is $\bar{a}_X \in \mathbb{R}^m$ such that $\varphi(\bar{a}_X, \bar{b}_s)$ true $\Leftrightarrow s \in X$. For $i = 1, 2, ..., n$, let $X_i = \{s \subseteq \{1, ..., n\} \mid i \in s\}$. We have $\bar{a}_{X_1}, \bar{a}_{X_2}, ..., \bar{a}_{X_n} \in \mathbb{R}^m$. Now for each $s \subseteq \{1, ..., n\}$, we have $\varphi(\bar{b}_s; \bar{a}_{X_i})$ true $\Leftrightarrow s \in X_i \Leftrightarrow i \in s$. $\qquad\square$

Now we can use these two lemmas to do a quick proof of Laskowski's theorem.

Proof.

$$C_\varphi \text{ is a VC-class} \Leftrightarrow \mathcal{V}(C_\varphi) = d \text{ for some } d \in \mathbb{N}$$
$$\Leftrightarrow \mathcal{I}(\psi) = d \text{ (by Lemma 1)}$$
$$\Rightarrow \mathcal{I}(\varphi) \leq 2^d \text{ (by Lemma 2)}$$
$$\Rightarrow \varphi \text{ does not have the independence property.}$$

Conversely,

$$\varphi \text{ does not have the independence property} \Leftrightarrow \mathcal{I}(\varphi) = n \text{ for some } n \in \mathbb{N}$$
$$\Rightarrow \mathcal{I}(\varphi) \leq 2^n \text{ (by Lemma 2)}$$
$$\Leftrightarrow \mathcal{V}(C_\varphi) \leq 2^n \text{ (by Lemma 1)}$$
$$\Rightarrow C_\varphi \text{ is a VC-class.}$$

$$\square$$

We say that the $\mathcal{L}$-structure $\mathfrak{R}$ has the *independence property* if there is a formula $\varphi(x; \bar{y})$ *with just the single variable* x that has the independence property with respect to $\mathfrak{R}$.

Applying model theoretic methods, Laskowski gives a clear combinatorial proof of the following theorem due to Shelah.

Theorem (Shelah [She71]). *An $\mathcal{L}$-structure $\mathfrak{R}$ has the independence property if and only if there is a formula $\varphi(\bar{x}; \bar{y})$ (in any number of x variables) that has the independence property with respect to $\mathfrak{R}$.*

This is a very hard theorem, and attests to Shelah's incredible combinatorial genius. Interestingly, if you look at the paper with Sauer's theorem, there is a footnote that quotes a referee as saying that these results had been proved earlier by Shellah.

Laskowski's theorem combined with the following result provides the link between *o*-minimality and VC-classes.

Proposition (Pillay–Steinhorn [PS86]). *O-minimal structures do not have the independence property.*

Theorem (Laskowski [Las92a]). *Let $\mathfrak{R} = (\mathbb{R}, <, ...)$ be o-minimal and let $S \subset \mathbb{R}^{k+m}$ be definable. Then the collection $\mathcal{C} = \{S_{\bar{x}} \mid \bar{x} \in \mathbb{R}^m\}$ is a VC-class.*

Thus any definable family of subsets in an *o*-minimal structure constitutes a VC-class. This theorem gives us a "black box" to generate an enormous variety of VC-classes.

Note that many structures are known not to have the independence property (by work of Shelah), and thus Laskowski's theorem provides significantly more examples of VC-classes. To illustrate, the field of complex numbers, $(\mathbb{C}, +, \cdot)$ does not have the independence property, and thus any definable family of sets in this structure is a VC-class.

5.2 Probably Approximately Correct (PAC) Learning

Begin with an *instance space* X that is supposed to represent all instances (or objects) in a learner's world. A *concept* c is a subset of X, which we can identify with a function $c : X \to \{0, 1\}$. A *concept class* $\mathcal{C}$ is a collection of concepts.

A *learning algorithm* for the concept class $\mathcal{C}$ is a function L which takes as input m-tuples $((x_1, c(x_1)), ..., (x_m, c(x_m)))$ for $m = 1, 2, ...$ and outputs hypothesis concepts $h \in \mathcal{C}$ that are consistent with the input. If X comes equipped with a probability distribution, then we can define the *error* of h to be $\text{err}(h) = P(h \Delta c)$.

The learning algorithm L is said to be PAC if for every $\varepsilon, \delta \in (0, 1)$ there is $m_L(\varepsilon, \delta)$ so that for *any* probability distribution P on X and any concept $c \in \mathcal{C}$, we have for all $m \geq m_{L(\varepsilon, \delta)}$ that

$$P\left(\{\bar{x} \in X^m \mid \mathrm{err}(L((x_i, c(x_i))_{i \leq m})) \leq \varepsilon\}\right) \geq 1 - \delta.$$

It can be shown that an algorithm that outputs a hypothesis concept h consistent with the sample data is PAC provided that $\mathcal{C}$ is a VC-class. Moreover, for given ϵ and δ, the number of sample points needed is, roughly speaking, proportional to the VC-dimension $\mathcal{V}(\mathcal{C})$.

5.3 Neural Networks

Macintyre–Sontag [MS93] and Karpinski–Macintyre [KM95] apply Laskowski's result and the uniform bounds available in o-minimal structures to answer questions about neural networks. The output in a sigmoidal neural network is the result of computing a quantifier-free formula whose atomic formulas have the form $\tau(\bar{x}, \bar{w}) > 0$ or $\tau(\bar{x}, \bar{w}) = 0$, where τ is built from polynomials and exp, $\bar{x}$ are input values, and $\bar{w}$ represent a tuple of programmable parameters. Varying the parameters gives rise to a definable family in an o-minimal structure and hence Laskowski's theorem applies, which tells us that it is possible to PAC learn the architecture of such a network.

The first results of Macintyre and Sontag applied Laskowski's theorem to prove finite VC-dimension. Using quantitative results of Khovanskii, Karpinski and Macintyre give an upper bound for the VC-dimension that is $O(m^4)$, where m is the number of weights. Koiran and Sontag [KS97] have established a quadratic lower bound (in the number of weights) for the VC-dimension.

Acknowledgments

We would like to thank Brendan Beare for his accurate reporting of Professor Steinhorn's five lectures on "Tame Topology and O-minimal Structures."

References

[AB91]   Adams, W., Brock, J.W.: Pareto optimality and antitrust policy: The old Chicago and the new learning. Southern Economic Journal **58**, 1–14 (1991)

[Afr67]   Afriat, S.: The construction of a utility function from demand data. International Economic Review **8**, 67–77 (1967)

[Afr72a]   Afriat, S.: Efficiency estimates of production functions. International Economic Review **13**, 568–598 (1972)

[Afr72b]   Afriat, S.: The theory of international comparison of real income and prices. In: Daly, J.D. (ed) International comparisons of prices and output. National Bureau of Economic Research, New York (1972)

[Afr77]   Afriat, S.: The price index. Cambridge University Press, London: (1977)

[Afr81]   Afriat, S.: On the constructability of consistent price indices between several periods simultaneously. In: Deaton, A. (ed) Essays in applied demand analysis. Cambridge University Press, Cambridge (1981)

[And86]   Anderson, R.M.: 'Almost' implies 'near'. Transaction of the American Mathematical Society **296**, 229–237 (1986)

[Ant98]   Anthony, M: Probabilistic generalization of functions and dimension-based uniform convergence results. Statistics and Computing **8**, 5–14 (1998)

[AB88]   Arnond, D.S., Buchberger, B.: Algorithms in real algebraic geometry. Academic Press, New York (1988)

[Arr51]   Arrow, K.: Social Choice and Individual Values. Yale University Press, New Haven (1951)

[AD54]   Arrow, K., Debreu, G.: The existence of an equilibrium for a competitive economy. Econometrica **22**, 265–290 (1954)

[AH71]   Arrow, K., Hahn: General Competitive Analysis. Holden-Day, San Francisco (1971)

[Bac94]   Backhouse, R. (ed): New Directions in Economic Methodology. Routledge Press, London and New York (1994)

[Bal88]   Balasko, Y.: Foundations of the Theory of General Equilibrium. Boston: Academic Press (1988)

[BPR03]   Basu, S., Pollack, R., Roy, M.-F.: Algorithms in Real Algebraic Geometry. Springer-Verlag, New York (2003)

194   References

[Bat57]   Bator, F.: The simple analytics of welfare maximization. American Economic Review **47**, 22–59. (1957)

[BR90]   Benedetti, R., Risler, J.-J.: Real Algebraic and Semi-algebraic Sets. Hermann, Paris (1990)

[Bis67]   Bishop: Foundations of Constructive Analysis. McGraw-Hill, New York (1967)

[Bla92]   Blaug, M.: The Methodology of Economics: Or How Economists Explain. Cambridge University Press, Cambridge (1992)

[BCSS98]   Blum, L., Cucker, F., M. Shub, Smale, S.: Complexity and Real Computation. Springer-Verlag, New York (1998)

[BZ93]   Blume, L., Zame, W.R.: The algebraic geometry of competitive equilibrium. In: Neuefeind, W. (ed) Essays in general equilibrium and international trade: In memoriam Trout Rader. Springer-Verlag, New York (1993)

[BEHW89]   Blumer, A., Ehrenfeucht, A., Haussler, D., and Warmuth, M.: Learnability and the Vapnik–Čherovenkis dimension. Journal of the AMC **36**, 929–965 (1989)

[BCR98]   Bochnak, J., Coste, M.,Roy, M.-F.: Real Algebraic Geometry. Springer-Verlag, Berlin (1998)

[Boh94]   Bohm, V.: The foundations of the theory of monopolistic competitive revisited. Journal of Economic Theory **63**, 208–218 (1994)

[Bor62]   Borch, K: Equilibrium in reinsurance markets. Econometrica **30**, 424–444 (1962)

[BPZ02]   Bossaerts, P., Plott, C., Zame, W.R.: Prices and portfolio choices in financial markets: Theory and experimental evidence. NBER Conference Seminar on General Theory, May 10–12, 2002 (2002)

[Bor98]   Borel, A.: Twenty-five years with Nicholas Bourbaki (1949–1973). Notices of American Mathematical Society, 373–380 (1998)

[BC07]   Brown, D.J., Calsamiglia, C.: The nonparametric approach to applied welfare analysis. Economic Theory **31**, 183–188 (2007)

[BH83]   Brown, D.J., Heal, G.M.: Marginal vs. average cost pricing in the presence of a public monopoly. American Economic Review **73**, 189–193 (1983)

[BK03]   Brown, D., Kannan, R.: Indeterminacy, nonparametric calibration and counterfactual equilibria. Cowles Foundation Discussion Paper 1426, Yale University, New Haven (2003)

[BK06]   Brown, D.J., Kannan, R.: Two algorithms for solving the Walrasian inequalities. Cowles Foundation Discussion Paper No. 1508R, Yale University (2006)

[BM93]   Brown, D.J., Matzkin, R.L.: Walrasian comparative statics. SITE Technical Report #85 (1993)

[BM95]   Brown, D.J., Matzkin, R.L.: Estimation of nonparametric functions in simultaneous equations models, with an application to consumer demand. Mimeo, Northwestern University (1995)

[BM96]   Brown, D.J., Matzkin, R.L.: Testable restrictions on the equilibrium manifold. Econometrica **64**, 1249–1262 (1996)

[BS00]   Brown, D.J., Shannon, C.: Uniqueness, stability, and comparative statics in rationalizable Walrasian markets. Econometrica **68**, 1528–1539 (2000)

[BW04]   Brown, D.J., Wood, G.A.: The social cost of monopoly." Cowles Foundation Discussion Paper 1466, Yale University (2004)

[Bro10]   Brouwer, L.E.J.: Über eineindeutige, stetige Transformationen von Flächen in sich. Mathematische Annalen **67**: 176–180 (1910)

[CRS04]   Carvajal, A., Ray, I., Snyder, S.: Equilibrium behavior in markets and games: Testable restrictions and identification. Journal of Mathematical Economics **40**, 1–40 (2004)

[CJ98]   Caviness, B.F., Johnson, J.R. (eds): Quantifier Elimination and Cylindrical Algebraic Decomposition. Springer-Verlag, Vienna (1998)

[Chi88]   Chiappori, A.: Rational household labor supply. Econometrica **56**, 63–89 (1988)

[CE98]   Chiappori, P.-A., Ekeland, I.: Restrictions on the equilibrium manifold: The differential approach. Working paper (1998)

[CEKP04]   Chiappori, P.-A., Ekeland, I., Kubler, F., Polemarchakis, H.: Testable implications of general equilibrium theory: A differentiable approach. Journal of Mathematical Economics **40**, 105–119 (2004).

[CR87]   Chiappori, A., Rochet, J.C.: Revealed preferences and differentiable demand. Econometrica **55**, 687–691 (1987)

[Col75]   Collins, G.E.: Quantifier elimination for the elementary theory of real closed fields by cylindral algebraic decomposition. In: Proceedings of the Second GI Conference on Automata Theory and Formal Languages. Springer-Verlag, New York 1975)

[CD96]   Constantinides, G.M., Duffie, D.: Asset pricing with heterogeneous consumers. Journal of Political Economy **104**, 219–240 (1996)

[Cor77]   Cornwall, R.R.: The concept of general equilibrium in a market economy with imperfectly competitive producers. Metroeconometrica **29**, 55–72. (1977)

[Dan00]   Dana, R.-A.: Stochastic dominance, dispersion and equilibrium asset pricing. Mimeo (2000)

[Dan63]   Dantzig, G.B.: Linear programming and extensions. Princeton University Press, Princeton (1963)

[DSW00]   Dawkins, C., Srinivasan, T.N., Whalley, J.: Calibration. In: Heckman, J.J., Leamer, E. (eds) Handbook of Econometrics, Vol. 5. Elsevier, New York (2000)

[Deb51]   Debreu, G.: The coefficient of resource utilization. Econometrica **19**, 273–292 (1951)

[Deb59]   Debreu, G.: Theory of Value. New York: Wiley (1959)

[Deb69]   Debreu, G.: Neighboring economic agents. La Decision. Paris: Presses du C.N.R.S., 85–90 (1969)

[Deb70]   Debreu, G.: Economies with a finite set of equilibria. Econometrica **38**, 387–392 (1970)

[Deb74]   Debreu, G.: Excess demand functions. Journal of Mathematical Economics **1**, 15–21 (1974)

[Deb86]   Debreu, G.: Mathematical Economics: Twenty Papers of Gerard Debreu. Cambridge University Press, Cambridge (1986)

[DM94]   den Haan, W.J., Marcet, A.: Accuracy in simulations. Review of Economic Studies **61**, 3–17 (1994)

[DV88]   Denef, J., van den Dries, L.: p-adic and real subanalytic sets. Annals of Mathematics **128**, 79–138 (1988)

[Dre74]   Drèze, J.H.: Investment under private ownership: Optimality, equilibrium and stability. In Drèze, J.H. (ed) Allocation under Uncertainty: Equilibrium and Optimality. John Wiley & Sons, New York, 129–165 (1974)

[DGMM94]   Duffie, D., Geanakoplos, J.D., Mas-Colell, A., McLennan, A.: Stationary Markov equilibria. Econometrica **62**, 745–782 (1994)

[DS86]   Duffie, D., Shafer, W.: Equilibrium in incomplete markets II. Journal of Mathematical Economy **15**, 199–216 (1986)

[EZ89]   Epstein, L., Zin, S.: Substitution, risk aversion and the temporal behavior of consumption and asset returns: A theoretical framework. Econometrica **57**, 937–969 (1989)

[GV04]   Gabrielov, A., Vorobjov, N.: Complexity of computation with Pfaffian and Noetherian functions. In: Normal Forms, Bifurcations and Finiteness Problems in Differential Equations. Kluwer, Dordrecht (2004)

[GVZ03]   Gabrielov, A., Vorobjov, N., Zell, T.: Betti numbers of semialgebraic and sub-Pfaffian sets. Journal London Mathematical Society **69**, 27–43 (2003)

[GJ95]   Goldberg, P., Jerrum, M.: Bounding the Vapnik–Čherovenkis dimension of concept classes parameterized by real numbers. Machine Learning **18**, 131–148 (1995)

[GP01]   Gul, F., Pesendorfer, W.: Temptation and self-control. Econometrica **69**, 1403–1436 (2001)

[HH96]   Hansen, L.P., Heckman, J.J.: The empirical foundation of calibration. Journal of Economic Perspectives **10**, 87–104 (1996)

[Har54]   Harberger, A.C.: Monopoly and resource allocation. American Economic Review Papers and Proceedings **44**, 77–87 (1954)

[Har12]   Hardy, G.H.: Properties of logarithmico-exponential functions. Proceedings of the London Mathematical Society **10**, 54–90 (1912)

[HPS00]   Haskell, D., A. Pillay, A., Steinhorn, C.: Overview. In: Haskell, D., Pillay, A., Steinhorn, C. (eds) Model Theory, Algebra, and Geometry (Mathematical Sciences Research Institute Publications, Vol. 39). Cambridge University Press, Cambridge (2000)

[HL96]   Heaton, J., Lucas, D.J.: Evaluating the effects of incomplete markets on risk sharing and asset pricing. Journal of Political Economy **104**, 443–487 (1996)

[Hel82]   Hellwig, M.: Rational expectations and the Markovian property of temporary equilibrium processes. Journal of Mathematical Economics **9**, 135–144 (1982)

[Hig96]   Higham, N.J.: Accuracy and Stability of Numerical Algorithms. SIAM, Philadelphia (1996)

[Hou50]   Houthakker, H.: Revealed preference and the utility function. Economica **17**, 1590-174 (1950)

[Hyl03]   Hylton, K.N.: Antitrust Law. Cambridge University Press, Cambridge (2003)

[II90]   Ingaro, B., Israel, G.: The invisible hand. MIT Press, Cambridge (1990)

[Ing90]   Ingram, B.F.: Equilibrium modeling of asset prices: Rationality versus rules of thumb. Journal of Business and Economic Statistics **8**, 115–125 (1990)

[JL03]   Jha, R., Longjam, I.S.: Structure of financial savings during Indian
           economic reform. Unpublished manuscript, Research School of Pacific
           and Asian Studies (2003)
[Jud92]   Judd, K.L.: Projection methods for solving aggregate growth models.
           Journal of Economic Theory **58**, 410–452 (1992)
[Jud97]   Judd, K.: Computational economics and economic theory: Comple-
           ments or substitutes? Journal of Economic Dynamics and Control **21**,
           907–942 (1997)
[Jud98]   Judd, K.L.: Numerical Methods in Economics. MIT Press, Cambridge
           (1998)
[Kak41]   Kakutani, S.: A generalization of Brouwer's fixed point theorem. Duke
           Mathematical Journal **7**, 457–459 (1941)
[KM97]   Karpinski, M., Macintyre, A.: Polynomial bounds for VC dimension of
           sigmoidal and general Pfaffian neural networks, Journal of Computing
           and System Sciences **54**, 169–176 (1997)
[KM95]   Karpinski, M., Macintyre, A.: Polynomial bounds for VC dimension
           of sigmoidal neural networks. In: Proceedings of the Twenty-seventh
           ACM Symposium on Theory of Computing. ACM, New York (1995)
[KV94]   Kearns, M., Vazirani, U.: An Introduction to Computational Learning
           Theory. MIT Press, Cambridge and London (1994)
[Kho80]   Khovanskii, A.: On a class of systems of transcendental equations. So-
           viet Math. Dokl. **22**, 762–765 (1980)
[Kho91]   Khovanskii, A.: Fewnomials. In: Translations of Mathematical Mono-
           graphs, Vol. 88. American Mathematical Society, Providence (1991)
[KK86]   Kirman, A., Koch, K.-J.: Market excess demand in exchange economies
           with collinear endowments. Review of Economic Studies **174**, 457–463
           (1986)
[KPS86]   Knight, J., Pillay, A., Steinhorn, C.: Definable sets in ordered struc-
           tures II. Transactions of the AMS **295**, 593–605 (1986)
[Koc96]   Kocherlakota, N.: The Equity premium: it's still a puzzle. Journal of
           Economic Literature 34, 42–71 (1996)
[Koi95]   Koiran, P.: Approximating the volume of definable sets. In: Proceedings
           of the 36th IEEE Symposium on Foundations of Computer Science.
           IEEE, New York (1995)
[KS97]   Koiran, P., Sontag, E.: Neural networks with quadratic VC dimension.
           Journal of Computing and System Sciences **54**, 190–198 (1997)
[Koo60]   Koopmans, T.C.: Stationary ordinal utility and impatience. Economet-
           rica **28**, 286–309 (1960)
[Kre04]   Krebs, T.: Consumption-based asset pricing with incomplete markets.
           Journal of Mathematical Economics **40**, 191–206 (2004)
[Kre90]   Kreps, D.M.: A Course in Microeconomic Theory. Princeton University
           Press, Princeton (1990)
[KP78]   Kreps, D.M., Porteus, E.L.: Temporal resolution of uncertainty and
           dynamic choice theory. Econometrica **46**, 185–195 (1978)
[Kub03]   Kubler, F.: Observable restrictions of general equilibrium models with
           financial markets. Journal of Economic Theory **110**, 137–153 (2003)
[Kub04]   Kubler, F.: Is intertemporal choice theory testable? Journal of Mathe-
           matical Economics **40**, 177–189 (2004)
[Kub07]   Kubler, F.: Approximate generalizations and computational experi-
           ments. Econometrica **75**, 967–992 (2007)

[KP04]   Kubler, F., Polemarchakis, H.M.: Stationary Markov equilibria in over-lapping generations. Economic Theory **24**, 623–643 (2004)

[KS05]   Kubler, F., Schmedders, K.: Approximate versus exact equilibria in dynamic economies. Econometrica **73**, 1205–1235 (2005)

[KS02]   Kubler, F., Schmedders, K.: Recursive equilibria with incomplete markets. Macroeconomic Dynamics **6**, 284–306 (2002)

[KS03]   Kubler, F., Schmedders, K.: Stationary equilibria in asset-pricing models with incomplete markets and collateral. Econometrica **71**, 1767–1793 (2003)

[KS07]   Kubler, F., Schmedders, K.: Competitive equilibria in semi-algebraic economies. PIER Working Paper No. 07-013 (2007)

[LP81]   Landes, W.M., Posner, R.A.: Market power in antitrust cases. Harvard Law Review **94**, 937–996 (1981)

[LM94]   Landsberger, M., Meilijson, I.: Co-monotone allocations, Bickel–Lehmann dispersion and the Arrow–Pratt measure of risk aversion, Annals of Operations Research **52**, 97–106 (1994)

[Las92a]   Laskowski, M.C.: Vapnik–Červonenkis classes of definable sets. Journal of the London Mathematical Society **245** 377–384 (1992)

[Las92b]   Laskowski, M.C.: Vapnik–Čherovenkis classes of definable sets. Journal of the London Mathematical Society **294**, 231–251 (1992)

[LEc02]   L'Ecuyer, P.: Random number generation. In: Gentle, J.E., Haerdle, W., Y. Mori, Y. (eds) Handbook of Computational Statistics. Springer-Verlag, New York (2004)

[LB07]   Lee, Y.A., Brown, D.J.: Competition, consumer welfare, and the social cost of monopoly. In: Collins, W.D. (ed) Issues in Competition Law and Policy. American Bar Association Books, Washington, DC (2007)

[Ler37]   Lerner, A.P.: The concept of monopoly and the measurement of monopoly power. Review of Economic Studies **1**, 157–175 (1937)

[Luc78]   Lucas, R.E.: Asset prices in an exchange economy., Econometrica **46**, 1429–1445 (1978)

[MM97]   Macintyre, A., Marker, D.: A failure of quantifier elimination. Revista Mathematics de la Universidad Computensa **10**, 209–216 (1997)

[MS93]   Macintyre, A., Sontag, E.: Finiteness results for sigmoidal "neural' networks. In: Proceedings of the 25th Annual ACM Symposium on Theory of Computing. Association for Computing Machinery, New York (1993)

[MW96]   Macintyre, A., Wilkie, A.: Schanuel's conjecture implies the decidability of real exponentiation. In: Odifreddi, P. (ed) Kreiseliana. A.K. Peters Ltd. (1996)

[Mac00]   Macpherson, D.: Notes on *o*-minimality and variations. In: Haskell, D., Pillay, A., Steinhorn, C. (eds) Model Theory, Algebra, and Geometry (Mathematical Sciences Research Institute Publications, Vol. 39). Cambridge University Press, Cambridge (2000)

[MQ96]   Magill, M.J.P., Quinzii, M.: Theory of incomplete markets. MIT Press, Cambridge, MA (1996)

[MSi93]   Maheswaran, S., Sims, C.: Empirical implications of arbitrage-free asset markets. In: P.C.B. Phillips (ed) Methods, models, and applications of econometrics: Essays in honor of A.R. Bergstrom. Blackwell, Oxford (1993)

[MPS05]   Mailath, G., Postlewaite, A., Samuelson, L.: Contemporaneous perfect epsilon equilibria. Games and Economic Behavior **53**, 126–140 (2005)

[Mal72]   Malinvaud, E.: Lectures on Microeconomic Theory. North–Holland, London (1972)

[Mar00]   Marker, D.: Introduction to model theory. In: Haskell, D., Pillay, A., Steinhorn, C. (eds) Model Theory, Algebra, and Geometry (Mathematical Sciences Research Institute Publications, Vol. 39). Cambridge University Press, Cambridge (2000)

[Mar02]   Marker, D.: Model Theory: An Introduction (Graduate Texts in Mathematics, Vol. 217). New York: Springer-Verlag, New York (2002)

[MS94]   Marker, D., Steinhorn, C.: Definable types in o-minimal theories. Journal of Symbolic Logic **59**, 185–198

[Mas77]   Mas-Colell, A.: On the equilibrium price set of an exchange economy. Journal of Mathematical Economics **4**, 117–126 (1977)

[Mas85]   Mas-Colell, A.: The Theory of General Economic Equilibrium. A Differentiable Approach. Cambridge University Press, New York (1985)

[MWG95]   Mas-Colell, Whinston, M.D., Green, J.R.: Microeconomic Theory. Oxford University Press, New York (1995)

[MR91]   Matzkin, R.L., Richter, M.K.: Testing strictly concave rationality. Journal of Economic Theory **53**, 287–303 (1991)

[Mil94]   Miller, C.: Exponentiation is hard to avoid. Proceedings of the Americal Mathematical Society **122**, 257–259 (1994)

[Mor00]   Morgan, T.D.: Cases and Materials on Modern Antitrust Law and Its Origin. West Group Press (2000)

[MR93]   Mourgues, M.H., Ressayre, J.-P.: Every real closed field has an integer part. Journal of Symbolic Logic, 641–647 (1993)

[Nac02]   Nachbar, J.: General equilibrium comparative statics. Econometrica **70**, 2065–2074 (2002)

[PS86]   Pillay, A., Steinhorn, C.: Definable sets in ordered structures I, Transactions of the AMS, **295**, 565–592 (1986)

[Pos75]   Posner, R.A.: The social costs of monopoly and regulation. Journal of Political Economy **83**, 807 (1975)

[Pos01]   Posner, R.A.: Antitrust Law. University of Chicago Press (2001)

[PS81]   Postlewaite, A., Schmeidler, D.: Approximate Walrasian equilibrium and nearby economies. International Economics Review **22**, 105–111 (1981)

[Qua98]   Quah, J.: The monotonicity of individual and market demand. Nuffield Working Paper Number 127 (1998)

[Res93]   Ressayre, J.-P.: Integer parts of real closed exponential fields. In: Clote, P., Krajiček (eds) Arithmetic, Proof Theory and Computational Complexity. Oxford University Press, Oxford (1993)

[RW99]   Richter, M.K., Wong, K.-C.: Non-computability of competitive equilibrium. Economic Theory **14**, 1–27 (1999)

[RW00]   Richter, M.K., Wong, K.-C.: Definable utility in o-minimal structures. Journal of Mathematical Economics **34**, 159–172 (2000)

[Rio96]   Rios-Rull, V.: Life-cycle economies and aggregate fluctuations. Review of Economic Studies **63**, 465–489 (1996)

[Roc87]   Rochet, J.C.: A necessary and sufficient condition for rationalizability in a quasilinear context. Journal of Mathematical Economics **16**, 191–200 (1987)

[Roc70]   Rockafellar, R.T.: Convex Analysis, Princeton University Press, Princeton (1970)

[Roj00]   Rojas, J.M.: Some speed-ups and speed limits in real algebraic geometry. Journal of Complexity **16**, 552–571 (2000)

[RSW03]   Rolin, J.-P., Speissegger, P., Wilkie, A.: Quasianalytic Denjoy-Carleman classes and o-minimality. Journal of the American Mathematical Society **16**, 751–777 (2003)

[Sam47]   Samuelson, P.A.: Foundations of Economic Analysis. Harvard University Press, Cambridge (1947)

[San00]   Santos, M.S.: Accuracy of numerical solutions using the Euler equation residuals. Econometrica **68**, 1377–1402 (2000)

[SP05]   Santos, M.S., Peralta-Alva, A.: Accuracy of simulation for stochastic dynamic models. Econometrica **73**, 1939–1976 (2005)

[SV98]   Santos, M.S., Vigo-Aguiar, J.: Analysis of a numerical dynamic programming algorithm applied to economic models. Econometrica **66**, 409–426 (1998)

[Sau72]   Sauer, N.: On the density of families of sets. Journal of Combinatorial Theory Series A **13**, 145–147 (1972)

[Sch91]   Schanuel, S.: Negative sets have Euler characteristic and dimension. In: Lecture Notes in Mathematics, Vol. 1488. Springer-Verlag, Berlin (1991)

[Sca60]   Scarf, H.E.: Some examples of global instability of the competitive equilibrium. International Economic Review **1**, 157–172 (1960)

[Sca67]   Scarf, H.: On the computation of equilibrium prices. In: Fellner, W. (ed) Ten Studies in the Tradition of Irving Fisher. Wiley, New York (1967)

[Sca73]   Scarf, H.E.: The computation of economic equilibrium. Yale University Press, New Haven (1973)

[She71]   Shelah, S.: The f.c.p., and superstability; model theoretic properties of formulas in first order theory. Annals of Mathematical Logic **3**, 271–362 (1971)

[Sho76]   Shoven, J.B.: The incidence and efficiency effects of taxes on income from capital. Journal of Political Economy **84**, 1261–1283 (1976)

[SW92]   Shoven, J., Whalley, J.: Shoven, J.: Applying General Equilibrium. Cambridge University Press, Cambridge (1992)

[Sil90]   Silberberg, E.: The Structure of Economics. McGraw Hill, New York (1990)

[Sim89]   Sims, C.A.: Solving nonlinear stochastic optimization and equilibrium problems backwards. Discussion Paper, University of Minnesota (1989)

[Sny04]   Snyder, S.K.: Observable implications of equilibrium behavior of finite data. Journal of Mathematical Economics **40**, 165–176 (2004)

[Spe03]   Speissegger, P.: O-minimal expansions of the real field. Preprint (2003)

[Sep99]   Speissegger, P.: The Pfaffian closure of an o-minimal structure. J. Reine Angew. Math. textbf508, 198–211 (1999)

[Str56]   Strotz, R.H.: Myopia and inconsistency in dynamic utility maximization. Review of Economic Studies **23**, 165–180 (1956)

[Stu02]   Sturmfels, B.: Solving Systems of Polynomial Equations. CBS Regional Conference Series in Mathematics, no. 97. American Mathematical Society, Providence (2002)

[Tar51]   Tarski, A.: A Decision Method for Elementary Algebra and Geometry, 2d ed. rev. Rand Corporation, Berkeley and Los Angeles (1951)

[Tho65]   Thom, R.: Sur l'homologie des variétés algébriques réelles. In Cairns, S.S. (ed) Differential and combinatorial topology: A Symposium in Honor of Marston. Princeton University Press, Princeton (1965)

[Uza62]   Uzawa H.: Walras' existence theorem and Brouwer's fixed-point theorem. Economic Studies Quarterly **8**, 59–62 (1962)

[Van86]   van den Dries, L.: A generalization of the Tarski–Seidenberg theorem, and some nondefinability results. Bulletin of the AMA **15**, 189–193 (1986)

[Van88]   van den Dries, L.: Alfred Tarski's elimination theory for real closed fields. Journal of Symbolic Logic **53**, 7–19 (1988)

[Van98]   van den Dries, L.: Tame topology and O-minimal structures In: London Mathematical Society Lecture Note Series, Vol. 248. Cambridge University Press, Cambridge (1998)

[Van00a]   van den Dries, L.: O-minimal structures and real analytic geometry. In: Current Developments in Mathematics 1998. International Press, Cambridge, MA (2000)

[Van00b]   van den Dries, L.: Classical model theory of fields. In: Haskell, D., Pillay, A., Steinhorn, C. (eds) Model Theory, Algebra, and Geometry (Mathematical Sciences Research Institute Publications, Vol. 39). Cambridge University Press, Cambridge (2000)

[VMM94]   van den Dries, L., Macintyre, A., Marker, D.: The elementary theory of restricted analytic fields with exposition. The Annals of Mathematics **140**, 183–204 (1994)

[VMM97]   van den Dries, L., Macintyre, A., Marker, D.: Logarithmic-exponential power series. Journal of the London Mathematical Society **56**, 417–434 (1997)

[VM94]   van den Dries, L., Miller, C.: On the real exponential field with restricted analytic functions. Israel Journal of Mathematics **85**, 19–56 (1994)

[VS98]   van den Dries, L., Speissegger, P.: The real field with convergent generalized power series. Translations of Mathematical Society **350**, 4377–4421 (1998)

[Var82]   Varian H.: The nonparametric approach to demand analysis. Econometrica **50**, 945–973 (1982)

[Var83]   Varian, H.: Varian, H.: Non-parametric tests of consumer behavior. Review of Economic Studies **50**, 99–110 (1983)

[Var84]   Varian, H.: The non-parametric approach to production analysis. Econometrica **52**, 579–597 (1984)

[Var92]   Varian, H.: Microeconomic Analysis, 3d ed. Norton, New York (1992)

[Van99]   van den Dries, L.: Tame Topology and O-Minimal Structures. Cambridge University Press, Cambridge (1999)

[VM44]   von Neumann, J., Morgenstern, O.: Theory of Games and Economic Behavior. Princeton University Press, Princeton (1944)

[Wal54]   Walras: Elements of Pure Economics. Allen and Unwin, London (1954)

[Wei85]   Weintraub, E.R.: General Equilibrium Analysis: Studies in Appraisal. Cambridge University Press, Cambridge (1985)

[Wil96]   Wilkie, A.J.: Model-completeness results for expansions of the real field by restricted Pfaffian functions and the exponential functions. Journal American Mathematical Society **9**, 1051–1094 (1996)

[Wil63]   Wilkinson, J.H.: Rounding Errors in Algebraic Processes. Prentice-Hall, Englewood Cliffs, NJ (1963)

[Wil84]   Wilkinson, J.H.: The perfidious polynomial. In: Golub, G.H. (ed) Studies in Mathematics, Vol. 24. Mathematical Association of America, Washington, DC (1984)

Lecture Notes in Economics and Mathematical Systems

For information about Vols. 1–519
please contact your bookseller or Springer-Verlag